STELLAR EVOLUTION

PLATE I. The Milky Way in Sagittarius. (Photograph by F. E. Ross.)

STELLAR EVOLUTION

AN EXPLORATION
FROM THE OBSERVATORY

By Otto Struve

1950

PRINCETON UNIVERSITY PRESS

PRINCETON, NEW JERSEY

PRINTED IN THE UNITED STATES OF AMERICA BY
THE COLONIAL PRESS INC., CLINTON, MASS.

TO

Henry Norris Russell

PREFACE

THE purpose of observational research in astrophysics is to present a unified picture of a series of phenomena and to explain it in terms of a theory or hypothesis. The temptation is always to accumulate more and more factual data and to delay the process of interpretation, because we rarely, if ever, feel satisfied that we have enough information to justify a generalization. It requires some outside stimulus to interrupt this trend and to direct the researcher's attention to the task of drawing conclusions from the available material.

The Vanuxem Lectures at Princeton in 1949 presented a favorable occasion for taking stock in one field of astrophysics—that of the origin and evolution of single stars and, more especially, of close double stars. This subject has been treated before by other workers, but since their observational basis was different from mine I thought that it would be interesting to present this subject in the light of my own experience at the telescope. I found the process an exhilarating one and I have thoroughly enjoyed writing this book. I am, however, conscious of the fact that of necessity there is a great deal of speculation in any attempt to discuss the evolution of the stars; the mind of a scientist trained to recognize new facts and to measure them shies away from the hollowness of weakly founded hypotheses or of various *ad hoc* assumptions. In some respects the observational astronomer knows too many facts to be entirely satisfied with almost any hypothesis: there are always some exceptional stars which defy him. The book is not an exhaustive survey and some of the hypotheses presented in it will be of value primarily in guiding future efforts.

In trying to discover the origin and the evolution of the stars the astronomer is confronted with only a snapshot of the galaxy as it is now. He is in the same position as an observer from a distant planet would be if he were given the task of describing the evolution of mankind after having seen the earth for a period of one second. He might record the existence of children, mature persons, and older men and women; and he might correctly conclude that there is evolution from young to old. But, as Professor E. Finlay-Freundlich of St. Andrews recently remarked, he might also conclude that since evolution means growth, a child will ultimately develop into an elephant! Erroneous conclusions of this kind can easily enter into our discussions and we must at all times be aware of the hypothetical nature of the evolutionary processes which we shall discuss.

The history of previous evolutionary hypotheses teaches us that most of them were wrong. Yet, they have contributed to our understanding of the universe and have, in almost every case, left a permanent imprint upon later hypotheses. I hope that the present effort will be useful in a similar manner. It is not intended as a final explanation of the development of the stars, but only as an attempt to try out certain ideas and see where they lead us. In doing so we shall encounter many weaknesses but also some interesting and unexpected connections.

In the theoretical aspects of this book I have been influenced by three recent publications: C. F. von Weizsaecker, "Zur Kosmogonie" (*Zs. f. Ap.*, 24, 181, 1947); A. Unsöld, "Kernphysik und Kosmologie" (*Zs. f. Ap.*, 24, 278, 1948); and V. Ambarzumian, *The Evolution of Stars and Astrophysics* (Armenian Academy of Sciences, Erevan, 1947). The theory of turbulence developed by Taylor, Prandtl, v. Weizsaecker, Heisenberg, Kolmogoroff, and Chandrasekhar has brought a new and remarkably fruitful element into astronomy. A critical review of the earlier hypothesis of von Weizsaecker, on the origin of the solar system, has been recently published by D. ter Haar ("Studies on the Origin of the Solar System," Det Kgl. Danske Videnskabernes Selskab, *Mat.-Fys. Medd.*, 25, No. 3, 1948). Two important Russian reviews, by V. G. Fessenkoff, "The Problem of Cosmogony in Contemporary Astronomy" (*Astronomical Journal of the Soviet Union*, 26, 67, 1949) and A. J. Massevich, "Stellar Evolution Accompanied by Corpuscular Radiation" (*Astronomical Journal of the Soviet Union*, 26, 207, 1949) closely parallel the results of our Chapter II. They arrived after this manuscript had been completed. An up-to-date summary of the nuclear theory of stellar energy has been presented by one of its originators, R. d'E. Atkinson, in the Halley Lecture for 1949 (*The Observatory*, 69, 161, 1949).

In keeping with the original purpose of the lectures, this book is intended not so much for astronomers as for physicists, chemists, geologists, and all those who are interested in the processes of evolution in the material universe. We cannot be sure that we understand all astronomical phenomena correctly. But we have, perhaps for the first time in history, a few fleeting glimpses of the truly magnificent and awe-inspiring panorama of creation. We are no longer limited to our own ability to reason out what went before and what will come next, but we can actually see, with our own eyes, how stars shed matter at their equators and produce gaseous rings and envelopes around them, thereby losing mass and rotational momentum; or how newly formed stars gather up dust from the clouds of diffuse interstellar matter in which they are embedded, and thereby gain mass, and probably rotational momentum.

The exposition is, for the most part, popular, and should present no great difficulties to the layman. But a few technical developments have been retained to whet the appetite of the reader and to show him how the results have actually been obtained. The presentation is usually historical in character and has been strongly influenced by the work of my colleagues at the Yerkes and McDonald Observatories.

Most of the illustrations have been taken from other publications, but a few, viz. No.'s 2, 4, 5, 6, 8, 9, 10, 11, 12, 13, 14, and 15 have been redrawn by Mr. Henry Horak, to whom I am indebted for this help. In the present form these diagrams are consistent with one another, but no effort has been made to change the luminosities to the bolometric scale where they were not already so given by the original authors. The diagrams for the Pleiades and Praesepe were originally given in the form of color-apparent magnitude plots. Mr. Horak has converted them into the conventional H-R form, in such a way as to obtain the best fit for the main sequence.

I am indebted to many astronomers for information, published and unpublished. I want to record especially my gratitude to Professor H. N. Russell for many valuable suggestions, to Dr. P. Swings for many thought-provoking discussions, and to Dr. B. Strömgren for an account of his present views concerning the problem of hydrogen content. Dr. E. Schatzman has directed my attention to certain inconsistencies between the results of observation and the theoretical path which a star might be expected to follow. Miss Alice Johnson has been responsible for most of the photographic work, and Mrs. Margaret Phillips has done the typing. Mr. Robert Hardil made a number of helpful suggestions. The *Astrophysical Journal* has furnished most of the illustrations. Other acknowledgments appear in the text and the legends of the figures.

OTTO STRUVE
*Andrew MacLeish Distinguished
Service Professor of Astrophysics*

Yerkes Observatory
University of Chicago
May 15, 1949

CONTENTS

STELLAR EVOLUTION

PROBING THE STARS'
CHEMICAL COMPOSITION

1. Introduction

PERHAPS the most remarkable property of the starry sky is the tremendous diversity among the apparent brightnesses of the individual stars. The brightest of the fixed stars, Sirius, is almost two billion times more luminous, as seen from the earth, than the faintest stars that can be photographed with our most powerful telescopes. We realize at once that this diversity in apparent luminosity must indicate a large range in stellar distances; and, in fact, one hundred years ago astronomers tried to derive these distances upon the assumption that the intrinsic luminosities of the stars are all approximately the same. But this assumption, necessary though it was when no better method was available, clearly led to some rather serious inconsistencies. For example, a cluster of stars, like the Pleiades, is from its shape and appearance clearly discerned as an aggregation of stars which are all at approximately the same distance from us. Such a cluster contains bright, blue stars which are easily seen with the naked eye (Alcyone has an apparent magnitude of 3.0, about 2.5 times fainter than the Pole star whose apparent magnitude is 2.0), while the faintest stars in the cluster, visible only in large telescopes, are red in color and are almost a million times fainter. This is an intrinsic difference in luminosity, not one produced by differences in distance. Other clusters permit us to extend the true range in intrinsic luminosity—among the known stars there are some which are a million times more luminous than the sun and there are others which are a million times less luminous, a range of almost 10^{12}.

At the same time, we have already commented upon the range in color among the stars of the Pleiades. This is evidently due to differences in the surface temperatures of the stars, so that in this respect, also, there is a large range in values among the building blocks of our galaxy, the Milky Way. The hottest stars may have surface temperatures of the order of 100,000°K, while the coolest stars are barely red-hot, 2,000°K at the most, which is less than the temperature of

3

an electric laboratory furnace and very much less than the temperature of the inside an atomic bomb in explosion.

We have also convincing evidence that the stars differ enormously in size. The infrared component of the double star Epsilon Aurigae is so large that traveling with an orbital velocity of 30 km/sec around its bright primary star it eclipses the latter for about 500 days! This means that the segment of the infrared star behind which the bright star passes during its eclipse has a length of

$$d = 30 \times 500 \times 24 \times 60 \times 60 \text{ km} = 2 \times 10^9 \text{ km}.$$

If the eclipse were central, d would be the diameter of the infrared star. In reality the eclipse is not central, and the plane of the orbit is not exactly in the line of sight. It turns out that the diameter of the infrared star is approximately 4×10^9 km, or 3,000 times the diameter of the sun. On the other hand, the smallest known stars, the white dwarfs, have diameters no larger than that of the earth, or even of the moon—400 times smaller than the sun. The range in radius, therefore, is of the order of 10^6.

As we shall see, the stars also differ greatly in mass, although the range is not as large as in radius or luminosity. The most massive stars exceed the sun by a factor of about fifty. The least massive are ten times lighter than the sun. Investigations under way at the Sproul Observatory of Swarthmore College, under the direction of P. van de Kamp, have made it probable that there are even less massive stars. Thus, K. Aa. Strand has announced the existence of a third component in the system 61 Cygni, with a mass of 0.016 that of the sun—only 16 times greater than that of Jupiter.

With all this diversity in the physical characteristics of the stars, the question naturally arises whether all possible combinations of intrinsic brightness, temperature, and radius, as well as of mass, are equally probable.

Let us imagine that we have plotted, in a three-dimensional model, the luminosities, radii, and masses as the three coordinates. Each star would be represented by a point in the space outlined by our three coordinates. If all combinations of values were equally probable the points would fill the volume uniformly. In reality, however, it turns out that many regions of the volume are empty. The points arrange themselves in several groups, or sheets, some of which are quite thin so that they resemble mathematical surfaces, while others are fairly thick, resembling three-dimensional volumes. Moreover, the density of the points within the surfaces and volumes is not the same. One narrow band which extends from small values of mass, radius, and luminosity, runs diagonally across the octant, clear through to the largest values of all three parameters and contains the vast majority of all the points. This band is called the main sequence. The sun belongs to it, and so do most of the familiar stars, Sirius, Vega, Procyon, etc. But even within the band the density of the points is not uniform. It is about one hundred times greater near the lower end, where mass, radius, and luminosity are small, than at the upper end.

This remarkable property of the stars to occur preferentially with certain definite combinations of their physical characteristics suggests that evolutionary trends have favored these, and no other, combinations. It is for this reason that whenever astronomers meet to discuss their problems they can be heard to repeat over and over again the mysterious letters L, $\mathfrak{M}$, and R. To the initiated they designate the three principal physical parameters which determine the individual character of a star: its luminosity, mass, and radius. Unfortunately, we cannot measure these quantities directly. Visual or photographic observations give us the apparent luminosity and not the intrinsic candlepower of the star. In order to find the latter we must know its distance and also the absorbing properties of the interstellar medium.

2. The Luminosities of the Stars

For the nearby stars the distances can be determined by means of a trigonometric procedure which makes use of the orbital motion of the earth around the sun. The usual method consists in photographing, through a large telescope, the star whose distance is to be determined, together with a field of very distant stars. The photograph is repeated half a year later when the earth in its motion around the sun has moved to a new position in space 186 million miles removed from the original position. This displacement of the point of observation causes a slight "parallactic" displacement of the nearby star with respect to the more distant stars of the galactic system. We measure the displacement in microns on the photographic plate and, knowing the focal length of the telescope, we can easily convert it into angular displacements on the apparent celestial sphere. For the nearest star, Alpha Centauri, the maximum displacement obtainable in this manner is two seconds of arc, or about 0.2 millimeter, in the focal plane of a very large telescope. This is the angle under which the diameter of the earth's orbit would be seen from the star. One-half this angle is called the parallax of the star. The distance itself, in kilometers or miles, follows immediately from the simple trigonometric solution of the isosceles triangle whose base is the diameter of the earth's orbit and whose apex is formed by the star. Because of the very great distances of all the fixed stars it is convenient to use as the unit of measurement a quantity designated as the parsec. This is the distance of a star from which the diameter of the earth's orbit would appear to subtend an angle of 2 seconds of arc, or the radius of the earth's orbit an angle of 1 second of arc. The parsec is equal to 3×10^{13} km. If a star has a parallax angle of 0.1 second of arc its distance is 10 parsecs or 3×10^{14} km, etc. Astronomers have decided to compare the intrinsic luminosities of the stars by assuming that they are all brought to a distance of 10 parsecs. Hence, if a star happens to be at that distance its apparent luminosity is said to be equal to its intrinsic or absolute luminosity; but if the real distance of a star is not 10 parsecs but, let us say, D parsecs, then the absolute luminosity exceeds the apparent luminosity by a factor of $(D/10)^2$, because the intensity of light is inversely proportional to the square of the distance. Thus, if the apparent luminosity is l the intrinsic luminosity would be

$$L = l(D/10)^2.$$

But astronomers frequently measure the luminosity of a star, apparent as well as intrinsic, on a logarithmic scale; and they describe these quantities as stellar magnitudes:

$$m = -2.5 \log l$$
$$M = -2.5 \log L.$$

The second expression can be transformed by means of the relation between L and l, into

$$M = -2.5 \log l - 5 \log D + 5 = m + 5 - 5 \log D.$$

For example, let us suppose that a star whose apparent luminosity corresponds to visual magnitude 6 has been found by the trigonometric method to be at a distance of 100 parsecs. If this same star could be brought to a distance of 10 parsecs the intensity of its light would be increased 10^2 or 100 times. Hence, the difference in magnitude is 2.5 log 100, so that δm is equal to 5. Our star will thus be 5 magnitudes brighter than we actually observe it to be, and since the magnitude scale is adjusted in such a way that brighter stars have smaller values of m the absolute magnitude of the star would be $+1$. We can verify this result by substituting 100 for D and computing the absolute magnitude from the formula $M = m + 5 - 5 \log D$.

In these computations we have neglected the effect of absorption in interstellar space. In reality the space between the stars is not completely transparent, and very distant stars appear somewhat fainter and at the same time redder than would be the case if space were transparent. A suitable correction can sometimes be introduced when the absorption is known, but in many cases this quantity is quite uncertain. A star may happen to be located in a very transparent region where the dimming of starlight may amount to only a few tenths of a magnitude per 1,000 parsecs. On the other hand, many stars are partly obscured by dark clouds of cosmic dust and gas. Some of these clouds are nearly opaque, so the stars behind them cannot be seen at all. Such clouds are present in the constellations Taurus, Ophiuchus, and others. On the average the absorption in interstellar space is approximately one magnitude per 1,000 parsecs; hence, the correct expression for the absolute magnitude is $M = m + 5 - 5 \log D - K$, where the quantity K is the photographic or visual absorption for the star under consideration.

One further convention must be explained before we can proceed and make use of the scale of absolute magnitudes. The stars are not all of the same color. Some are blue, others are reddish. Hence, the apparent magnitudes will not be the same if we measure the light of a star with a visual photometer or with a blue-sensitive photographic plate. Similarly, the absolute magnitudes will not be the same if we compute them with the formula given above. Hence, it is necessary to distinguish between visual absolute magnitudes and photographic absolute magnitudes. Both scales, and some others, are in use. For example, there are photoelectric absolute magnitudes and red or even infrared absolute magnitudes. Yet, for our purposes the really significant physical quantity is the luminosity of a star corresponding to its entire radiation. This quantity cannot be directly measured because the earth's atmosphere absorbs all radiations shorter than about wave length 3,000 Å and most radiations longer than about wave length 10,000 Å; but we know from various investigations that it is possible, at least as a first approximation, to consider the radiation of a star as that of a black body whose temperature is equal to the effective temperature of the star. Thus, if we have some measure of the star's surface temperature we can make corrections for the amount of light which is not accessible to our observations and we can express the luminosity in terms of the total radiation in all wave lengths. This kind of absolute magnitude is described as the bolometric magnitude. The corrections which must be made in passing from one of the other systems to the bolometric system have been tabulated and are also available in graphical form. We are never really concerned with the zero-point of this correction, since it is immaterial whether we call the absolute magnitude of the sun $+4.8$ or some other value. Hence, we have adopted the convention of calling the bolometric correction of the sun zero. This is convenient because it so happens that in solar-type stars most of the radiation is concentrated in the visual region of the spectrum. For hotter stars the maximum is displaced towards the unobservable ultraviolet; for cooler stars it is in the unobservable infrared. Hence, the bolometric magnitudes of stars of other types are brighter than their visual or photographic magnitudes—by as much as 2 magnitudes for the O and M stars, respectively.

The bolometric absolute magnitude is particularly interesting because it is related in a very simple way to the temperature of the star and its radius. According to the law of Stefan-Boltzmann the total radiation of a square centimeter of the surface of a black body in all directions is σT^4 where σ is a numerical constant accurately known from physical experiments and theory. If the radius of the star is R in centimeters then the total radiation is $4\pi R^2 \sigma T^4$ and the bolometric absolute magnitude

$$M_b = -2.5 \log 4\pi R^2 \sigma T^4.$$

In many cases it is not possible to determine accurately the distance of a star and thus to derive the absolute bolometric magnitude. But occasionally we can observe groups of stars, such as those located in globular star clusters or in open galactic clusters, like the Pleiades, in which we know from the spatial arrangement of the individual stars that they are all approximately at the same distance from us. If we measure the apparent magnitudes of these stars and take them relative to a fixed standard, we obtain very accurate measures of differences in intrinsic luminosity or absolute magnitude and these values are often preferable to those obtained by applying individual corrections for the distances of the stars. But, of course, it must be remembered that we can discuss in such cases only relative absolute magnitudes and not their real values.

3. The Radii of the Stars; Their Temperatures and Spectra

The next most important physical quantity for each star is its radius. This again is a quantity that cannot be determined directly, except in the case of a very small number of giant and supergiant stars which can be measured by means of a very large interferometer. Such measures are difficult at best and they are available for such a small number of objects that we cannot make extensive use of them. But they confirm values of the radius which are computed from the relation between the intrinsic luminosity and the surface brightness of the star. As we have already seen, the absolute bolometric magnitude

$$M_b = -2.5 \log 4\pi R^2 \sigma T^4.$$

Hence, if the absolute bolometric magnitude is known and if in addition we know the temperature of the star, we can compute its radius. We already know how M_b can be determined. Our next question is how to determine the temperature. In the case of the sun it is possible to measure, by means of a suitable instrument, the amount of radiant energy which is received by each square centimeter of a surface placed at right angles to the incoming rays. This quantity is somewhat less than the radiation which would have been received if there were no absorption in the earth's atmosphere. But by measuring this quantity at different elevations of the sun above the horizon we can vary the path which the light has traveled through the air and then extrapolate the measured quantities for zero thickness of air. The result is usually given per minute of time and is expressed in calories. It is called the solar constant and amounts to 1.90 calories per square centimeter per minute or 1.33×10^6 ergs per square centimeter per second. The quantity which we would like to know is the amount of radiant energy which is emitted in all directions per second of time by each square centimeter of the sun's surface. If the distance of the earth from the sun is r and if the radius of the sun is R then the total radiated energy at the sun's surface would be $(r/R)^2$ times greater than the solar constant. The quantity r/R is the reciprocal of the angle in radians under which the radius of the sun, 16 minutes of arc, is seen from the earth. Hence, r/R equals approximately 215, and the solar radiation per square centimeter per second

$$\sigma T^4 = 6 \times 10^{10} \text{ erg/cm}^2/\text{sec.}$$

The constant of the Stefan-Boltzmann law $\sigma = 5.8 \times 10^{-5}$, expressed in centimeter-gram-second units. Introducing this numerical value, we find that the surface temperature of the sun, which we describe as the effective temperature, is $T_e = 5{,}700°$.

In the case of the fixed stars it is usually not possible to determine the effective temperatures because we cannot measure the actual amount of radiant energy coming from these stars. Hence, it is necessary to employ other, less direct, methods of inferring their temperatures.

The most useful method of determining the temperatures of the fixed stars is to measure as accurately as is possible the distribution of the intensity at different wave lengths in the continuous spectra. We know from detailed studies of this sort that the energy distributions of the different stars tend to resemble the distribution predicted by Planck's radiation law

$$I_\lambda = \frac{c_1}{\lambda^5} \frac{1}{e^{c_2/\lambda T} - 1}.$$

Although we now know that the use of the black-body energy distribution, which is implied by the adoption of Planck's function, is only a crude approximation, the results are fairly good. It is desirable in practice to measure the energy within the continuous spectrum of a long interval in wave length. In this manner we can derive empirical curves which can then be matched with a series of theoretical curves computed for different values of the temperature. The value of T which provides the best fit is described as the color temperature of the star. Sometimes it is not even necessary to use a spectrograph and to make measurements within very narrow intervals of wave length. With the help of colored filters, photographic and especially photoelectric determinations of the apparent magnitudes of the stars can be used in order to determine the differences of magnitude obtained in different wave lengths. These quantities are called the color indices of the stars and they can be accurately calibrated in terms of the color temperature.

Finally, there are several excellent spectroscopic methods which enable us to determine the excitation temperature of a star's atmosphere and its ionization temperature. The former is derived by means of intensity measurements of spectral lines of a single atom or ion. If we have measured a spectral line originating from an excited level whose potential is E_s in two stars which we shall designate by the prime and double prime superscripts, then the Boltzmann relation for each star gives us the population in level s with respect to the ground level 1:

$$\log n'_s = \log n'_1 - E_s \frac{5040}{T'}$$

$$\log n''_s = \log n''_1 - E_s \frac{5040}{T''}.$$

We have neglected two small corrections which measure the statistical weights of each spectroscopic term. The values n'_1 and n''_1 are not known but we can eliminate them if we measure a pair of spectral lines, one at level s and the other at level t. We then find by subtraction

$$\log \frac{n'_s}{n''_s} - \log \frac{n'_t}{n''_t} = 5040(E_t - E_s)\left(\frac{1}{T'} - \frac{1}{T''}\right).$$

Hence, if the excitation potentials are known and if, in addition, we know the temperature of one of the two stars we can, in principle, compute the temperature of the other star. In practice, however, it is not always easy to express the ratios n'_s/n''_s and n'_t/n''_t in terms of the measured absorptions within the spectral lines. The difficulty which here arises is one that is always present when we try to interpret the absorption spectrum of a star. Only in the case of very faint spectral lines are the two ratios of the populations in the relevant atomic states directly proportional to the measured absorptions within the spectral lines. For stronger lines this direct proportionality breaks down, and it is necessary to have recourse to a method involving the construction of a so-called curve of growth which represents in a rather complicated manner the relation between the numbers of absorbing atoms and the corresponding intensities of the resulting absorption lines. Granting that such a curve can be constructed, there is no further difficulty in determining the excitation temperature of a star with respect to some standard such as the sun. This method has been used a great deal in recent years. It has often led to temperatures which are somewhat lower than those obtained by the color measurements, but the two quantities are closely related to one another and it is always possible to convert the one into the other. The reason why they are not identical is that the spectral lines are produced at a somewhat higher level in the atmosphere of the star than the continuous spectrum. Consequently, the temperatures indicated by the lines tend to be a little lower than the temperatures indicated by the continuous spectrum.

There is also in use a scale of temperatures which rests upon the application of M. N. Saha's ionization equation. This formula connects the ratio of the numbers of ionized to un-ionized atoms with the temperature of the medium and the pressure of the free electrons in it. Using the same logarithmic form in which we described the Boltzmann relation, we can write the ionization equation in the following form:

$$\log \frac{n^+}{n} \, p_e = -I \frac{5040}{T} + \frac{5}{2} \log T - 0.48.$$

In this case we designate by n^+ the number of ionized atoms and by n the number of un-ionized atoms; p_e is the electron pressure and I is the ionization potential. We have neglected a small numerical constant which should be added to the right-hand side of the equation and which depends upon the statistical weights of the various states of the ionized and unionized atoms. The form of the ionization equation is considerably more complicated than that of the Boltzmann relation. The latter involves only the temperature, T, the former depends not only upon the temperature but also upon the electron pressure. In some rather important cases we can determine the electron pressure by means of other considerations. For example, from the measurements of the Stark effect of the hydrogen and helium lines we can often find directly the electron densities and hence the corresponding pressures. When that has been done it is necessary only to measure the ratio n^+/n by comparing the lines of an un-ionized and an ionized atom. For example, we might measure the intensities of the lines H and K of Ca II and of the resonance line of Ca I λ 4226. If we have carried out this measurement in a single star, the ionization equation does not directly lead to the determination of T. But if we know, in addition, the transition probabilities corresponding to the two sets of lines, then the ratio n^+/n can be immediately determined and this leads to the determination of T. If the transition probabilities are not accurately known—and unfortunately that happens in a great majority of the astronomically important atoms—then we must resort to a different method, which consists in measuring the same pair of lines in two different stars. For example, we could measure a line of Fe II and a line of Fe I in the sun and in some other star, such as Procyon. By taking differences, as we did in the case of the Boltzmann formula, we eliminate the uncertainties of the transition probabilities and thus find T.

But we must be careful not to underestimate the importance of the electron pressure. For stars which are dwarfs and are located on the main sequence of the Hertzsprung-Russell diagram, R. H. Fowler and E. A. Milne long ago found empirically that p_e remains approximately constant. Making the assumption that this constancy is a characteristic phenomenon along the entire main sequence and is equal to the electron pressure in the sun, namely approximately 100 dynes per square centimeter (or 10^{-4} atmospheres), they derived a sequence of temperatures which has been described as the ionization temperature scale of the stars. In a more recent discussion, A. Pannekoek has derived the ionization temperature scale separately for the dwarfs and for the giants. For each group he used a constant value of the acceleration of gravity, rather than a constant value of the electron pressure. He adopted $\log g = 4.4$ for the main-sequence stars and $\log g = 2.4$ for the giants. Fortunately, we can test the accuracy of a temperature scale determined by one of the indirect methods by computing the effective temperatures from the luminosities and radii of several eclipsing variables. As we have already remarked several times, the luminosity $L = 4\pi R^2 \sigma T_e^4$. If the luminosity and the radius are expressed in terms of the sun we have

$$ L = R^2 \left(\frac{T_e}{T_e^\odot} \right)^4 . $$

In the case of an eclipsing variable the radii can be determined in terms of the distance between the centers of the two stars, as we had sketched it in the case of Epsilon Aurigae on page 4; but if, in addition, we know this separation in kilometers, which is always the case when a complete velocity-curve of both components of the binary is available, then we have all the data necessary to compute the effective temperatures. In a general way, we can say that the results of the various methods agree reasonably well, and a reliable temperature scale is now available for all stars except those of the earliest spectral types and a few small groups of very peculiar objects.

In place of the effective temperature it is often convenient to use an observational characteristic which can be directly determined from the spectra of the stars. This characteristic is described as the spectral type and it is indicated by a capital letter, O, B, A, F, G, K, M, N, R, or S. The order of the letters is not important: it arose historically from the manner in which the separate classes were discovered at the Harvard Observatory. It is essentially based upon the ionization in the atmosphere of the star and is therefore a measure not only of the temperature but also of the pressure in the star's atmosphere. However, in actual practice the spectral types were determined by means of the intensities of the more conspicuous absorption lines; and since these intensities depend upon the temperature and the pressure in a slightly different manner, a compromise has been effected so that the overwhelming majority of the stars can be arranged in two sequences, the giants and the dwarfs. Within each sequence the intensities of the spectral absorption lines change in a smooth manner so that for any given set of intensities the corresponding spectral type, and hence the temperature, may be inferred. Because of the lower pressure in the atmosphere of a giant star the ionization in its atmosphere is higher than in that of a dwarf. But in assigning a giant to a certain spectral type we have attempted to smooth out such differences. For example, we have used approximately similar ratios of intensities for giants and for dwarfs. As a consequence the giants have a somewhat lower temperature for any given spectral type than the dwarfs. The exact relation between temperature and spectral type has been carefully investigated so that for any normal dwarf or giant it is sufficient to determine the spectral type in order to know its temperature. We must remember, however, that there are a few unusual stars within our galaxy in which the ionization is anomalous and we cannot rely upon it to give us a dependable estimate of the temperature. The greatest uncertainty affects the Wolf-Rayet stars and the nuclei of some planetary nebulae. The spectra of these objects have strong emission lines of atoms in high stages of ionization. A method originally proposed by H. Zanstra and D. H. Menzel, and developed by the former, gives the temperature by assuming that the ultraviolet radiation of the star beyond the Lyman limit at λ 911 is completely absorbed in a nebular shell and is ultimately radiated by the latter in the form of Balmer emission lines. A modification of Zanstra's method by K. Wurm and T. L. Page has given values of the order of 200,000°K for some objects. Yet, these objects are not characterized by absorption lines that suggest a correspondingly high degree of ionization. For example, some Wolf-Rayet stars have the higher members of the Balmer series in absorption. The ionization is not high enough to completely ionize all H atoms. Perhaps we must be prepared to recognize the possibility of high ionization in a shell without a correspondingly high temperature of the exciting stellar radiation. This is illustrated by the solar corona in which emission lines of [Fe xiv] are found, despite the fact that the continuous radiation of the sun computed with $T = 6000°$ is exceedingly weak at those wave lengths which are required to remove 13 electrons from the atom of iron. Table I is sufficient for our purpose. It gives the approximate temperature of a star from type O to type M8. For stars of type G0, and cooler, we make a distinction between the giants and the members of the main sequence, or dwarfs, while for stars of earlier type a single scale is sufficient.

TABLE I

THE STELLAR TEMPERATURE SCALE

Spectral Type	Main Sequence, Temperature	Giants, Temperature
O5	30,000°K	—
B0	21,000°K	—
B5	14,000°K	—
A0	10,600°K	—
A5	8,200°K	—
F0	7,100°K	—
F5	6,300°K	—
G0	5,750°K	5,300°K
G5	5,400°K	4,500°K
K0	4,900°K	4,000°K
K5	4,300°K	3,200°K
M0	3,400°K	3,000°K
M2	2,870°K	2,810°K
M8	—	1,780°K

4. The Masses of the Stars

Our last concern is with the masses of the stars. This quantity is available only if the star is a member of a binary system and if orbital motion within the system has been measured.

The only source of information which we have for the determination of the masses of the stars is obtained from the observations of double stars, visual as well as spectroscopic. All of these determinations are based upon the application of Kepler's third law, which reads as follows: "The squares of the periods of the planets are proportional to the cubes of their mean distances from the sun." If we designate by t_1 and t_2 the periods of two planets in the solar system and by a_1 and a_2 their mean distances from the sun, then we can write

$$t_1{}^2 : t_2{}^2 = a_1{}^3 : a_2{}^3.$$

From this law Newton inferred that the forces acting on the planets are inversely proportional to the squares of their distances and directly proportional to their masses. If we are now concerned with a pair of stars forming a binary system and assume that their motions are also determined by Newton's law of gravitation, then by means of Kepler's third law we can immediately write

$$\frac{\mathfrak{M}_1 + \mathfrak{M}_2}{sun + earth} = \frac{A^3}{P^2},$$

where $\mathfrak{M}_1$ and $\mathfrak{M}_2$ are the masses of the two stellar components, A is the distance between the two stars expressed in astronomical units, and P is the period of the binary system in years. Since the mass of the earth is negligible in comparison to the mass of the sun, we can simply write

$$\frac{\mathfrak{M}_1 + \mathfrak{M}_2}{\mathfrak{M}_\odot} = \frac{a^3}{\pi^3 P^2},$$

where we have inserted in place of A the ratio a/π and where π is the parallax of the star. By means of this expression it is easy to determine the total masses of visual double stars if we know the orbital elements and the distances. In some cases we can also determine the mass ratios of the components. This, however, is possible only if we can refer the motions of the two components to their center of gravity, instead of referring the fainter component to the brighter component, as is usually done. For example, on photographic plates obtained with a large telescope, it is often possible to measure the motions of the two stellar components with respect to a field of faint stars forming the background. In this manner we can determine the apparent ellipses of each component separately, and from the ratio of the axes of these two ellipses we immediately find the ratio of the masses.

Another very interesting method depends upon the observation of spectroscopic binaries. In such a system we observe, by means of the Doppler effect, the velocity with which one component or sometimes both components revolve with regard to the observer. If the two components of the spectroscopic binary are sufficiently bright to record their spectra simultaneously then we observe double lines at certain phases in the orbit. These double lines can be measured with respect to the comparison spectrum produced at the telescope, and from the amplitude of the oscillation we determine the velocity in the orbit. If we assume for a moment that the orbit is circular, then the orbital velocity, usually described by the letter K, is given by the expression

$$K_1 = \frac{2\pi a_1}{P},$$

where a_1 is the radius of the orbit of the component under consideration with respect to the center of gravity and P is the period. In this
case we usually express K in km/sec, so that a_1 must also be given in
kilometers and P must be given in seconds. If P is given in days, our
expression becomes

$$K_1 = \frac{a_1}{13{,}750\ P}.$$

Similarly, for the other component we obtain

$$K_2 = \frac{a_2}{13{,}750\ P}.$$

However, we must remember that we do not usually observe a system
whose orbital plane is exactly coincident with the line of sight. Hence,
we do not actually measure the velocity in the orbit, but only its
projection upon the line of sight. This makes it necessary to introduce
into our expressions a quantity depending upon the angle which the
plane of the orbit makes with the line of sight, and we arrive in this
manner at the two expressions:

$$K_1 = \frac{a_1 \sin i}{13{,}750\ P}$$

and

$$K_2 = \frac{a_2 \sin i}{13{,}750\ P}.$$

The ratio of the two velocities is equal to the ratio of the two masses:

$$K_2/K_1 = \mathfrak{M}_1/\mathfrak{M}_2.$$

If we now designate by the letter a the mean distance between the
centers of the two stars, it is easily shown that

$$\frac{a_1}{a} = \frac{\mathfrak{M}_2}{\mathfrak{M}_1 + \mathfrak{M}_2} \quad \text{and} \quad \frac{a_2}{a} = \frac{\mathfrak{M}_1}{\mathfrak{M}_1 + \mathfrak{M}_2}.$$

Again applying Kepler's law, as in the case of a visual binary system,
and expressing the masses of the two stars in terms of the mass of the
sun, we have

$$\mathfrak{M}_1 + \mathfrak{M}_2 = \frac{a^3}{P^2},$$

where a is expressed in astronomical units and P in years. If we observe in the spectrum only a single component, then the best that we
can do is to determine not the ratio a^3/P^2 but the ratio

$$\frac{(a_1 \sin i)^3}{P^2} = (\mathfrak{M}_1 + \mathfrak{M}_2) \frac{\mathfrak{M}_2^3 \sin^3 i}{(\mathfrak{M}_1 + \mathfrak{M}_2)^3} = \frac{\mathfrak{M}_2^3 \sin^3 i}{(\mathfrak{M}_1 + \mathfrak{M}_2)^2},$$

and this can be easily expressed in terms of the orbital velocity K_1 by means of the expression which we had derived previously. However, if we have in the spectrum the lines of both components, then we have a second expression of the same kind involving a_2 with the respective masses. It is then easy to see that we can determine the quantity

$$(\mathfrak{M}_1 + \mathfrak{M}_2) \sin^3 i = \frac{(a_1 + a_2)^3 \sin^3 i}{P^2},$$

and since the mass ratio is also known from the ratio K_2/K_1 we can determine separately the quantities

$$\mathfrak{M}_1 \sin^3 i$$

$$\mathfrak{M}_2 \sin^3 i.$$

If the orbits are eccentric, then a is the semi-major axis of the orbit, and the expressions of K_1 and K_2 contain the additional factor

$$(1 - e^2)^{-1/2}.$$

From spectroscopic observations alone it is not possible to determine the inclination i. However, many spectroscopic binaries happen to be at the same time eclipsing variables. That is, they are so oriented that one star periodically obscures the other. In the case of eclipsing binaries it is possible to determine the inclination i, thereby giving us the individual masses $\mathfrak{M}_1$ and $\mathfrak{M}_2$. It is by this method that the masses of a considerable number of binary systems have been determined, and the mass-luminosity relation as we now know it is entirely based upon information obtained from double stars.

The only other method which has been applied toward the determination of the masses of the stars depends upon the measurement of the gravitational red shift of the spectral lines of several stars. This method, for example, could be used for the determination of the mass of the companion of Sirius. It depends upon the fact that the gravitational red shift is proportional to the ratio $\mathfrak{M}/R$, where both the mass and the radius are expressed in terms of the sun's mass and the sun's radius. For the sun, the gravitational red shift is equal to 0.6 km/sec. For example, if we know that the red shift in a star is about 20 km/sec we can infer that the ratio $\mathfrak{M}/R$ in terms of the corresponding value for the sun is approximately 33. In order to determine the mass we must know the radius. This is the case if we know the surface brightness of the star and the total luminosity. For the companion of Sirius the mass obtained in this manner agrees well with the mass obtained by considerations of Kepler's third law.

R. J. Trumpler has attempted to use this method in the case of several hot stars which are members of galactic star clusters. He had found from observations made at the Lick Observatory that some O-type stars which happen to be members of clusters give radial velocities considerably in excess of the radial velocities obtained from other members of the same clusters. On the average, for quite a number of these galactic clusters, he obtained a red shift of the order of about 14 km/sec. It is true that the members of the cluster which are cooler and therefore do not show a large red shift will still be influenced to some extent by the gravitational effect. Hence, Trumpler corrected the measures for what he supposed to be a reasonable value of the red shift of the comparison stars and then proceeded to evaluate for each hot star of spectral type O the corresponding value of the ratio $\mathfrak{M}/R$. He was able to estimate the radii of the stars by means of the absolute magnitudes which he knew from the distances of the clusters and from the apparent magnitudes of these particular stars. He also knew, of course, their approximate temperatures because he had photographed the spectra of the stars and had estimated their spectral types. Hence, he knew the amount of radiation which they were emitting per square centimeter of their surfaces. Dividing the total luminosity by the radiation per square centimeter he determined the surface areas of the O-type stars, and thus their radii. In this manner he found values which ranged from about 100 times to about 300 or 400 times the mass of the sun. However, it is now believed that these excessive masses are not physically real. There is no question concerning the accuracy of Trumpler's observational data. The red shift undoubtedly is real, but as Trumpler had already realized when he announced his results, the interpretation of the red shift as a gravitational phenomenon involved an uncertainty which he was not at that time able to test. In the meantime, several additional pieces of information have come to light.

In 1948, J. A. Pearce at the Dominion Astrophysical Observatory in Victoria, B.C., determined the orbit of one of Trumpler's massive stars, HD 215835, which had been found to be a spectroscopic binary with a period of a little more than two days and with the lines of both stars showing during the elongations. The orbital velocity of the more massive component is about 262 km/sec, while the orbital velocity of the less massive component is 321 km/sec. The eccentricity is small, and for all practical purposes we can consider the motion as occurring in a circle. The result is:

$$\mathfrak{M}_1 \sin^3 i = 23\mathfrak{M}_\odot$$

$$\mathfrak{M}_2 \sin^3 i = 19\mathfrak{M}_\odot.$$

These masses are quite normal and are not consistent with the large values which Trumpler had derived under the assumption that these stars are single objects and that they are influenced by the effect of gravitation. A somewhat similar conclusion could have been drawn much earlier from the orbit of the spectroscopic binary Tau Canis Majoris. This system has a period of 154 days and the orbital velocity of the brighter component is approximately 50 km/sec as measured with respect to the center of gravity. The spectrum of the fainter component is not visible, consequently only the rather complicated mass function involving the masses of both components can be found:

$$\frac{\mathfrak{M}_2^3 \sin^3 i}{(\mathfrak{M}_1 + \mathfrak{M}_2)^2} = 1.9\mathfrak{M}_\odot.$$

If we designate the mass ratio $\alpha = \mathfrak{M}_1/\mathfrak{M}_2$, we can write this expression also in the form

$$\mathfrak{M}_2 \sin^3 i = 1.9(1 + \alpha)^2 \, \mathfrak{M}_\odot.$$

We have no observational information concerning the value of the mass ratio α, but if $\alpha = 2$ the mass function would give us

$$\mathfrak{M}_2 \sin^3 i = 17\mathfrak{M}_\odot,$$

$$\mathfrak{M}_1 \sin^3 i = 34\mathfrak{M}_\odot.$$

These are again quite normal stellar masses. Of course, it is possible that α is larger than 2, but this is not probable in the case of an O-type spectroscopic binary which is not known to be an eclipsing variable. Tau Canis Majoris belongs to the galactic star cluster NGC 2362. The velocity of the system is $+43$ km/sec, while that of the other stars in the cluster is $+34$ km/sec. The red shift is, therefore, $+9$ km/sec with respect to the comparison stars, or $+10$ km/sec with respect to an absolute standard. From this value Trumpler derived a ratio $\mathfrak{M}/R = 15.6$ times the corresponding value for the sun. The radius of the star is about 20 times that of the sun; therefore the mass, according to the gravitational explanation of the red shift, would have to be 300 times greater than that of the sun. It does not seem possible to reconcile this value with the perfectly normal masses resulting from the mass function, despite the fact that the inclination is not known.

In order to make this argument even stronger, we examined the spectra of some of the massive Trumpler stars in galactic clusters, especially that of 15 S Monocerotis, a member of the cluster NGC 2264. For this star Trumpler derived from the red shift a mass of 180 times the mass of the sun. Yet its spectrum is quite normal and resembles that of the average O-type star in the galaxy for which much smaller masses had previously been determined. We conclude that Trumpler's gravitational masses are not real and that the explanation of the red shift shown by the spectral lines must be different. Perhaps the relatively broad lines which we often observe in the spectra of these stars do not give us strictly the radial velocity of the center of mass of the star, but they contain a component of the velocity of gaseous masses which resemble streams or prominences. It is suggestive in this connection that the prominences of the sun have predominantly an inward motion. The atoms must be raised to high elevations in a greatly ionized condition where they produce no visible light and probably no absorption. They become visible as they fall back into the sun, and an outside observer would record, predominantly, motions of recession. This would appear to him as a red shift. In the case of the sun the prominences are not large enough, nor optically thick enough, to produce appreciable absorption lines, but it is entirely possible that in the hotter stars such absorption lines may be formed in the vast filamentary envelopes of prominences which we believe may exist in the outer regions of the atmospheres of hot stars.

The largest mass of any star for which we have observations is that of the spectroscopic binary, HD 47129. This system was discovered by J. S. Plaskett at Victoria on December 16, 1921. He found that its spectrum of type O has double absorption lines and that its period is 14.414 days. The semi-amplitude of the primary star was found to be 206 km/sec and that of the secondary star 247 km/sec. With these elements Plaskett computed the mass functions $\mathfrak{M}_1 \sin^3 i = 76\mathfrak{M}_\odot$ and $\mathfrak{M}_2 \sin^3 i = 63\mathfrak{M}_\odot$. The total mass of the system is, thus, approximately 139 times that of the sun; this is a minimum value because the star is not an eclipsing binary and the inclination is not known. It may not even be very close to 90°, in which case the mass of the system must be considerably larger. However, recent observations made at the McDonald Observatory have shown that there exist in the spectrum of Plaskett's star physical processes which had not been found in the earlier Victoria observations. For example, the intensities of the component lines undergo large variations. Especially the fainter component is very different in intensity at different times. It is never strong when the velocity of the primary is one of approach, and it is always fairly strong, sometimes even very strong, when the velocity of the primary is positive. The variations are, however, not entirely consistent and they may not even be exactly periodic. For example, it was found that in different cycles the intensities may be quite different. Sometimes the fainter component is almost as strong as the primary component, and yet, even during the same cycle, one or two nights later, the fainter component may be almost absent. Such variations in the intensities of the fainter component throw doubt upon the interpretation of Plaskett's measurements. Moreover, the McDonald Observatory results indicate that the radial velocity of the fainter component seems to be displaced by about 100 km/sec toward smaller values, algebraically, than are consistent with the velocity of the system. For example, when the primary component has a velocity of −200 km/sec, the velocity of the fainter, secondary component is of the order of +150 km/sec; but when the velocity of the primary component is approximately +200 km/sec the velocity of the secondary component is approximately −250 km/sec. This effect is systematic, but it is complicated by erratic variations from cycle to cycle and even from night to night. The most obvious way to explain the general shift of about −100 km/sec for the fainter component of HD 47129 is to suppose that it is surrounded by an atmosphere which is expanding so that we observe something in the nature of a P Cygni phenomenon. However, there are obstacles to this interpretation. It is not even quite certain that what we measure as the fainter component of HD 47129 is really a part of the secondary star. We must not forget that we observe an absorption line of varying intensity and of varying displacement. The erratic variations in displacement show that we do not measure the true velocity of the fainter component. We must be measuring, at least in part, the velocity of a stream of gas within the system, and there is no certainty that this stream of gas is projected in front of the fainter component. It could, for example, be projected in front of the brighter component and still give us a displacement roughly in the observed direction.

However, this is probably too far-fetched; for the present we shall suppose that we observe a stream of gas in front of the secondary component. A number of other cases of spectroscopic binaries are known in which we also observe similar inconsistencies. For example, recent observations of the massive system AO Cassiopeiae reveal phenomena which in many respects resemble those in HD 47129. The spectral lines are also visible in pairs, and in the early work at the Mount Wilson Observatory by W. S. Adams and G. E. Strömberg, and, later, in the work at Victoria by J. A. Pearce, the velocity curves of the two components were determined separately. The intensities of the secondary component undergo large changes. There is again a systematic variation, such that the secondary is stronger when the primary is receding. This was already noticed in the early Mount Wilson work. It was not commented upon by Pearce and we do not know whether there had been a change or whether Pearce had not noticed this variation, but there are also erratic changes, certainly from cycle to cycle, and probably within each cycle. The

period of this star is about 3.5 days. Hence, it is not possible to observe each cycle in great detail. On the other hand, we observe a
great many cycles in a relatively short interval of time and the impression is that the variations from cycle to cycle are conspicuous.

Then there are other early-type spectroscopic binaries in which
these phenomena, though present, are not very conspicuous. For
example, in Alpha Virginis, whose spectral type is early B, the intensities of the lines of the fainter component undergo variations which
are similar to, though smaller in range than, the variations observed
in the two O-type spectroscopic binaries. All these variations are
well known at the present time. They were first discovered at the
Harvard Observatory by S. Bailey and were later confirmed by Miss
Antonia C. Maury and Miss Annie J. Cannon.

These complications show that it is difficult to arrive at really reliable determinations of large stellar masses. In the case of less massive
stars the velocity-curves are usually not complicated by extraneous
phenomena and the values are quite dependable.

5. The Mass-Luminosity Relation

We are now ready to examine the three-dimensional relation be-
tween the quantities $\mathfrak{M}$, L, and R. In practice, it is convenient to
plot the stars, not in the form of a three-dimensional model, but in
two separate diagrams, one showing the relation between $\mathfrak{M}$ and L
and the other that between R and L. Using the formula $L = 4\pi R^2 \sigma T^4$
we can substitute for the (R,L) diagram one relating the temperature,
T, and the luminosity, L. The first is called the mass-luminosity
diagram, the second the radius-luminosity diagram and the third
the Hertzsprung-Russell or the temperature-luminosity diagram.

We start with the mass-luminosity diagram and show in Figure 1 a
plot of all stars made by G. P. Kuiper in 1938 for which reliable
masses and luminosities were available.* We notice that the dots do
not cover the entire area of the diagram uniformly, but are mostly
arranged in a fairly narrow band running from stars with small
masses and low luminosities to stars with large masses and large
luminosities. The existence of such a relationship was first established
in 1911 by J. Halm, from the observational data. It was discussed
by H. N. Russell in 1913 and by E. Hertzsprung in 1918. But the
modern form of the relation is due to A. S. Eddington (Plate II),
whose name is usually associated with it. Eddington also gave the
first theoretical explanation of this relation.

With the exception of the white dwarfs, like the companion of
Sirius, the stars follow fairly closely the relation

* We have omitted the Trumpler stars.

$$L = c\mathfrak{M}^{3.5}.$$

But the white dwarfs do not obey this formula; they are much too faint, intrinsically, considering that their masses are sometimes as large as the sun's. In evaluating the mass-luminosity relation we must remember that the information depends entirely upon binaries; we must not expect of necessity that the binaries and the single stars are identical in regard to the mass-luminosity relation. This is an important point when we consider the problem of the origin of the binaries. If, as we shall discuss later, the binaries have originated through some process of fission from an early-type parent star, then it is entirely possible that the components of the binary have not

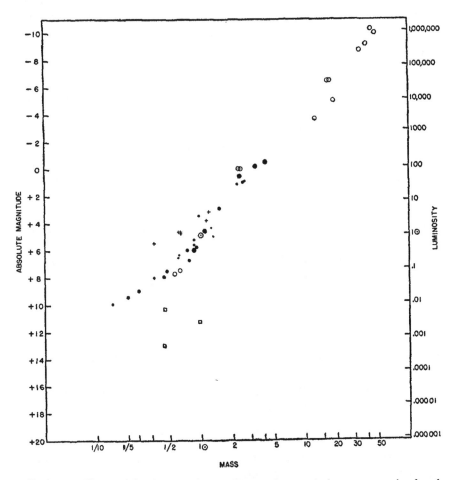

FIGURE 1. The mass-luminosity relation. Dots and open circles represent visual and spectroscopic binaries, each component being shown separately. Crosses represent several visual binaries in the cluster of the Hyades. Squares represent the white dwarfs.

entirely adjusted themselves and are in a state which is somewhat different from that of the undisturbed parent body. For example, we might easily find that these stars lie above the main sequence. This, in fact, seems to be the case for some of the very massive O-type binaries which are at the same time unusually luminous. The reason why we do not observe a discrepancy in the mass-luminosity relation is that we have no other stars of similar luminosity upon which to base our empirical mass-luminosity curve. Similarly, there is some reason to think that certain A- and B-type binaries have sub-giant secondaries of lower temperature. These secondaries are anomalous with respect to mass and luminosity, the latter being unusually large. The O-type systems which we are discussing now are probably stars which have divided into approximately equal masses. We would not expect to have quite as violent a disturbance in the case of such a division as we would have in the case of a division with a mass ratio departing greatly from one, but even so there is no certainty that the components were not somewhat disturbed and were not made to depart appreciably from the state which the parent star originally possessed. But this uncertainty does not exist when we are concerned with the visual binaries, or even with spectroscopic binaries, in which the separation between the components is so large as to preclude any disturbances either of one component by the other, or of each component as the result of a catastrophic change at the time when the binary was formed. There is no complete certainty about the manner in which the double stars were formed (see Chapter III), but all investigators agree that the pairs with large separations cannot be very young. They must have adjusted themselves completely to the conditions appropriate for their respective values of $\mathfrak{M}$, L, and R.

The most complete discussion of all available material on the masses of double stars was published in 1940 by H. N. Russell and Charlotte E. Moore. Their conclusions agree closely with those of Kuiper: there is a simple linear relation between the quantities log $\mathfrak{M}^{1/3}$ and $[M_{bol} + 2 \log T/5200]$. The term involving the temperature is of importance from theoretical considerations; it represents a small correction to the theoretical mass-luminosity relation which arises because L is not entirely independent of R or T, although it depends much more strongly upon $\mathfrak{M}$. This correction is closely correlated with the luminosity, so that we can write

$$2 \log T/5200 = -0.145(M_{bol} - 5).$$

This makes it possible to represent the mass-luminosity curve as an expression involving only L and $\mathfrak{M}$:

$$L = \text{const. } \mathfrak{M}^{3.82},$$

which is similar to Kuiper's result. There is no indication of a systematic difference between different groups of double stars. For example, the eclipsing variables, which cover the entire range of the curve, agree closely with the visual pairs, which cover only about one-half the range, namely, that of the less massive stars.

But it must not be thought that all stars fall exactly upon the mean curve. Even aside from the white dwarfs there are some which depart more or less seriously from it. Kuiper has called attention to six visual double stars in the cluster of the Hyades for which he finds luminosities in excess of those which are consistent with the curve. On the average these six stars are about three or four times more luminous than are stars of comparable mass on the curve.

Another famous departure is recorded for the star Zeta Herculis A. According to Kuiper its mass is almost identical with that of the sun, but its radius is twice, and its luminosity about four times, that of the sun. According to a recent determination by R. G. Hall, Jr., these quantities are somewhat changed so that, according to the best available data, it may now be taken to fall approximately 1 magnitude above the main sequence. These departures tend to spread the mass-luminosity curve into a band whose width may be several magnitudes along the side of M_{bol}. We shall see that this spread can be explained by differences in the chemical compositions of the stars: those with large hydrogen content have smaller luminosities than those which contain little hydrogen. Russell and Moore have found similar departures.

With respect to the empirical mass-luminosity relation, we can depend at the present time upon that part of the curve which extends from absolute bolometric magnitude $+10$, corresponding to a mass of about one-tenth that of the sun, and reaches a bolometric magnitude of about -10, corresponding to a mass of about 30 or 40 times the mass of the sun.

Recent investigations have made it abundantly clear that the components of some spectroscopic binaries seem to violate the mass-luminosity relation and form a group, or groups, of objects for which there exist few counterparts among the components of visual double stars, or among those single stars whose spectra have been investigated sufficiently for the determination of their surface gravities. However, these results are scattered throughout the astrophysical literature and are not easily accessible. In fact, it appears not to be generally known that the departures from the mass-luminosity relation may, in some cases, be truly enormous. Since the mass-luminosity relation has been derived almost entirely by means of data on visual and spectroscopic binaries and since this relation, in its accepted form, shows but a small amount of scatter, the large discrepancies must be confined to systems which, for one reason or another, were not made use of in deriving the relation.

R. M. Petrie at Victoria has for some years been engaged in the spectrophotometric determination of the relative luminosities, ($l = L_2/L_1$) of double-lined spectroscopic binaries. His most recent discussion, presented at the 1946 Harvard meeting of the American Astronomical Society, gave the results for a large number of systems ranging between $l = 0.04$ and $l = 1.00$. When his observed results are converted into a mass-luminosity relation, the great majority of objects agree reasonably well with the conventional curve, and the conclusion is justified that, with few exceptions, both components of the double-lined binaries follow quite closely the accepted relation.

But it must be remembered that the double-lined binaries in the first place furnished us with the observational data upon which rests a substantial portion of the mass-luminosity relation. Petrie's result, therefore, means that as a class the double-lined binaries yield the same relation as do the well-known systems that were originally chosen by Eddington, and, more recently, by G. P. Kuiper and by H. N. Russell and Miss Moore, in their discussions.

There exists, however, even among the double-lined binaries, a special group which seriously violates the mass-luminosity relation. One member of this group, i Bootis, was included in Petrie's work and gave a large discrepancy. The rest are also all W Ursae Majoris systems. Their mass-ratios have been determined from the observed values of K_1 and K_2, and are listed in Table XI.

The mean value of the mass-ratio, $\alpha = \mathfrak{M}_1/\mathfrak{M}_2 = 2$. For ordinary double-lined spectroscopic binaries $\alpha \approx 1.25$. The spectral types of the components of the W Ursae Majoris systems are nearly alike and their luminosities differ only slightly, since both sets of absorption lines are easily visible. In AH Virginis the individual masses of the components, if inserted in Kuiper's mass-luminosity relation, lead to a predicted difference $\Delta M = 3$ magnitudes. But the spectrographic intensities of the two sets of absorption lines and the light curve agree in giving $L_1 = 0.65$, $L_2 = 0.35$ and $\Delta M = 2.5 \log 0.65/0.35 = 0.7$ magnitude. This value must be close to the truth. That derived from the mass-luminosity relation is clearly incorrect. We conclude that one or both components violate the mass-luminosity relation, the discrepancy in this case being about 2 magnitudes.

The values of K_2 and K_1 in Table XI are uncorrected for the reflection effect. If this correction is introduced, K_2 and K_1 become larger, but K_2 more so than K_1. The result is an inappreciable change in α, and the discrepancy remains almost the same. It is tempting to attribute it to the close proximity of the two stars. Probably they have a common envelope through which an adjustment in the outer layers takes place which renders the components spectroscopically and photometrically more nearly alike than would be consistent with the mass-luminosity relation.

Our next group of systems which depart from the mass-luminosity relation are the Algol-type stars. These systems usually show only the spectrum of the smaller, brighter, and hotter component, outside of eclipse. The spectrum of the larger, fainter, and cooler component can often be observed during the total eclipse. In a few cases (U Sagittae, U Cephei) the slope of the velocity-curve of the fainter component has been derived during the total phase and in these cases the mass-ratios are known.

Since α is known individually for very few Algol stars, we must proceed statistically. The mass-function is

$$f(\mathfrak{M}) = \frac{\mathfrak{M}_2^3}{(\mathfrak{M}_1 + \mathfrak{M}_2)^2} \sin^3 i = \frac{\mathfrak{M}_2}{(1 + \alpha)^2} \sin^3 i = cPK_1^3(1 - e^2)^{3/2}.$$

If the ordinary spectroscopic binaries are plotted separately for each spectral class, with K_1 as the ordinate and $\log P$ as the abscissa, they fill an area limited at the bottom by the horizontal axis, at the top by a curve of the form $K_1 = f_1 P^{-1/3}$, and on the left side by a vertical straight line which is different for each spectral class and which corresponds to the smallest distance between the centers, $r_1 + r_2$, which a binary can have. If all systems of a given spectral class had the same values of $\mathfrak{M}_2$ and α, the vertical distribution of the points in each diagram would be caused solely by differences in i (and in e, but the latter can be allowed for). The eclipsing variables would then be located near the upper limiting curve.

In reality, when the Algol-type variables are plotted on the same diagrams, using the spectral class of the brighter component, we find that they form a group limited at the top by a similar curve, $K_1 = f_2 P^{-1/3}$, but the constants are quite different:

$$f_1/f_2 = 3.$$

This must mean that the Algol systems differ systematically from the ordinary spectroscopic binaries, in mass or in mass-ratio, or in both. The spectra of the brighter components of the Algol systems are usually normal for each spectral class. It is reasonable to suppose that their masses are also normal. If we make this assumption, we find that the mass-ratio is, on the average, about $\bar{\alpha} = 5$ and the mass of the fainter component is smaller than would have been expected from the giant or subgiant G-type or K-type spectra of these stars, or from their photometric values of r_2 which, in conjunction with the spectrographically determined values of $a_1 \sin i$ and the mass-ratios, give radii of the order of several times the radius of the sun.

These results are representative for an average Algol-type binary whose principal eclipse has a depth of $A_1 \geqslant 1.4$ magnitudes. In such systems the fainter component is so much weaker than the brighter component that there can be no blending of the spectral lines and hence no vitiation of the observed semi-amplitudes of the velocity curves of the more massive components, K_1.

There are individual stars for which α is even larger and $\mathfrak{M}_2$ smaller. Remarkable in this respect is XZ Sagittarii, investigated in 1945 at the McDonald Observatory by J. Sahade.

He found $K_1 = 11.5$ km/sec and $P = 3.28$ days. Hence,

$$f(\mathfrak{M}) = \frac{\mathfrak{M}_2}{(1 + \alpha)^2} \sin^3 i = 0.0005 \mathfrak{M}_\odot.$$

From the photometric solution we know that the ratio of the luminosities of the components is $L_b/L_f = 10$. The mass function gives the following sets of values:

$\mathfrak{M}_2$	α	$\mathfrak{M}_1$
$1.0 \mathfrak{M}_\odot$	44	$44 \mathfrak{M}_\odot$
0.16	17	3
0.02	5	0.1

The spectrum of the brighter component is entirely normal for a main-sequence star of class A. Its mass must be somewhere near $3 \mathfrak{M}_\odot$. Hence, $\alpha = 17$ and $\mathfrak{M}_2 = 0.16 \mathfrak{M}_\odot$. The spectrum of the fainter component was observed during totality and is approximately G.

If we substitute the masses of the two components of XZ Sagittarii into the empirical mass-luminosity relation, we obtain:

A3 component; $\mathfrak{M}_1 = 3 \mathfrak{M}_\odot$; $M_{bol} = + 0.7$; $M_{vis} = + 1.2$

G component; $\mathfrak{M}_2 = 0.16 \mathfrak{M}_\odot$; $M_{bol} = +12.8$; $M_{vis} = +12.8$.

The difference would be:

$$\Delta M_{vis} = 11.6 \text{ mag.}$$

In reality, however,

$$\Delta M_{vis} = 2.5 \log 10 = 2.5 \text{ mag.}$$

Hence, the companion is almost 10,000 times more luminous than would be consistent with the mass-luminosity relation! The very fact that we can easily photograph the G-type spectrum during the totality without prohibitively long exposures shows that the real value of ΔM is small.

In 1948 Sahade reobserved this system at the Bosque Alegre station of the Cordoba Observatory. The new series gave $K_1 = 23$ km/sec and $f(\mathfrak{M}) = 0.004 \mathfrak{M}_\odot$. The discrepancy is due to the difficulty of determining the amplitude of the velocity-curve, which is very small in both series of observations. With these new data we find that $\mathfrak{M}_2 = 0.35\mathfrak{M}_\odot$ and $\alpha = 8.5$ when $\mathfrak{M}_1 = 3\mathfrak{M}_\odot$. Sahade concludes that in all probability α lies between 8 and 16, and that the fainter component of XZ Sagittarii contradicts the mass-luminosity relation.

The discrepancy is enormous and would be regarded with suspicion if it represented an isolated case. Actually, it is not at all unique. As far back as in 1913, F. C. Jordan found that the eclipsing variable R Canis Majoris has $K_1 = 29$ km/sec and $P = 1.136$ days, so that $f(\mathfrak{M}) = 0.0027\mathfrak{M}_\odot$. The primary component is a normal star of class A9. Hence, Jordan concluded that its mass should be at least $\mathfrak{M}_1 = 1\mathfrak{M}_\odot$, and this would require $\alpha = 6.3$ and $\mathfrak{M}_2 = 0.16\mathfrak{M}_\odot$.

F. B. Wood has rediscussed the problem of R Canis Majoris. With the help of B. W. Sitterly's spectrographic elements and his own photometric observations he concluded that the mass-ratio would have to be as large as $\alpha = 10$ in order to give a normal mass for the brighter component, $\mathfrak{M} = 2.06\mathfrak{M}_\odot$, in which case the fainter component would have a mass of $\mathfrak{M}_2 = 0.21\mathfrak{M}_\odot$. As in the case of XZ Sagittarii the true luminosity of the fainter component, as derived from the light-curve, is thousands of times greater than would be consistent with the mass-luminosity relation.

If, contrary to the evidence of the spectrum, we should assume that not the mass of the primary, but that of the secondary, is normal for its spectral type and therefore is consistent with the mass-luminosity relation, then, as Wood has shown, the mass-ratio would have to be about $\alpha = 40$ and $\mathfrak{M}_1 = 144\mathfrak{M}_\odot$. These values are so unreasonable that we must discard them in favor of the abnormally small mass of $\mathfrak{M}_2 = 0.21\mathfrak{M}_\odot$. The important thing here is that it is impossible to reconcile the masses of both components with their spectral types and with the mass-luminosity relation.

Wood has given important reasons for believing that in reality α cannot be larger than about 5. Otherwise, part of the mass of the secondary would be located outside the Jacobian limiting surface of zero velocity having a double point where the two loops touch. In this case both components would be completely abnormal from the standpoint of the mass-luminosity relation:

principal component: $\mathfrak{M}_1 = 0.4\mathfrak{M}_\odot$; $M_{bol} = +\ 9.0$

secondary component: $\mathfrak{M}_2 = 0.06\mathfrak{M}_\odot$; $M_{bol} = +16.8$.

We have no reason to doubt the very simple conclusions based upon the spectrographic determination of $f(\mathfrak{M})$. We must recognize that at least the fainter components of some Algol-type binaries are stars of a kind not often recognized in the past. Since they do not usually occur among visual double stars, it is possible that they are abnormal only because they have not had time to recover from the catastrophe which led to the formation of the binary and to adjust themselves to conditions appropriate for their masses. But H. N. Russell has suggested that if the star Zeta Herculis A were set revolving around Sirius, it could give total eclipses of the Algol type. It would carry us too far to speculate further on this subject, except to say that if the adjustment is not now complete there is no stability, and hence not even a certainty that the present masses of the components are safe from dissipation.

Except for relatively minor features, the spectra of the G-type and K-type secondaries are not abnormal. They usually suggest a giant or a subgiant character, and this is consistent with their large photometric radii.

It is possible to go one step further, and examine statistically the distribution of the stars in a diagram relating K_1 and P (Fig. 22) for a given spectral class of the brighter component. We have a relation between K_1 and sin i of the form

$$K_1 P^{1/3}(1 - e^2)^{1/2} = \mathfrak{M}_1^{1/3}\alpha^{-1/3}(1 + \alpha)^{-2/3} \sin i$$

The frequency distribution of sin i with i is known: it shows a steep rise to maximum at $i = 90°$. Hence, if $\mathfrak{M}_1$ and α are constants, the distribution of the quantity on the left-hand side of our expression must also have a strong maximum for large values of K_1: the points in Fig. 22 should be strongly concentrated towards the upper limiting curve. Observational selection favors the discovery of systems with large K_1 and therefore accentuates the effect. In reality, there is no such concentration. This shows that the values $\mathfrak{M}_1$ and α cannot both be constants. To derive the true frequency distribution of $\mathfrak{M}_1$ and α we must solve an integral equation, or proceed, as Chandrasekhar and Münch have shown, to determine the moments of the observed distribution and from these derive the moments of the true distribution. The rigorous mathematical treatment is, however, not possible because we cannot accurately correct the observations for the probability of discovery. Moreover, we cannot separate the distribution of $\mathfrak{M}_1$ from that of α. Nevertheless, if we assume that the distribution of $\mathfrak{M}_1$ in the entire material does not differ greatly from that observed in double-lined binaries, we reach the conclusion that α shows a large dispersion: values of the order of 20 are fairly frequent, and values of the order of 100 are not unreasonable. In all probability, the masses of the companions in binary systems can have all values from $\mathfrak{M}_1$ down to those of planets. But it is only when these secondaries are abnormally luminous that we observe their spectra.

The third group of discordant systems occurs near the upper end of the mass-luminosity relation—among the stars of very large mass. The accepted mass-luminosity relation in this region rests upon a few well-known O-type and early B-type systems with double lines which are at the same time eclipsing variables. Among these systems, AO Cassiopeiae = HD 1337 = Boss 46 and UW Canis Majoris = 29 Canis Majoris play an important role. According to Kuiper's curve these stars, which have individual masses of the order of $40\mathfrak{M}_\odot$, have absolute bolometric magnitudes of the order of -10. There may be some doubt about the use of 29 Canis Majoris, because its fainter component is not visible on McDonald Observatory spectrograms. But even if, for the sake of caution, we should disregard this fainter component, which had led to $\alpha = 1.33$, the mass-function is so large, $f(\mathfrak{M}) = 4.57$, that we can be reasonably sure that $\mathfrak{M}_1 \geqslant \mathfrak{M}_2 \geqslant 20\mathfrak{M}_\odot$.

Much more important is the information which we derive from HD 698. It is seriously off the curve. J. A. Pearce measured double lines and found $f(\mathfrak{M}) = 3.6\mathfrak{M}_\odot$, $\alpha = 2.5$. This gives directly $\mathfrak{M}_1 \sin^3 i = 113\mathfrak{M}_\odot$; $\mathfrak{M}_2 \sin^3 i = 45\mathfrak{M}_\odot$. No eclipses have been observed. The mass-luminosity relation would require at least $M_1 \sim M_2 \sim -8$ (*bol*). But the spectrum, classified by Pearce as B9 sek, is that of a main-sequence star, because the H lines have broad wings produced by Stark effect. This would give a visual absolute magnitude of the order -1. The discrepancy is not as large as in the case of the fainter components of the Algol stars. Yet, it is large enough to be disturbing, since in this case it affects mostly the brighter component whose spectrum is easily photographed.

In the case of close binary systems, departures from the mass-luminosity curve do not usually indicate differences in chemical composition, as they do in distant pairs, such as Zeta Herculis. It is probable that we are here concerned with a phenomenon whose origin is related to the fact that the components are almost in contact. These unusual cases are relatively rare, and they do not alter the conclusions previously drawn with regard to the mass-luminosity relation: the great majority of the stars fall within the limits of the band outlined in the investigations of Kuiper and of Russell and Moore.

6. The Hertzsprung-Russell Diagram

About forty years ago E. Hertzsprung and H. N. Russell independently recognized that there are very few blue stars of low luminosity, and that the red stars consist of two distinct groups: those of low luminosity, similar to the sun or to the visual double star 61 Cygni, and those of high luminosity, similar to Antares or Betelgeuze. They designated the former as "dwarfs" and the latter as "giants." In 1911 Hertzsprung began plotting the luminosities of stars in several galactic clusters, the Pleiades, the Hyades, etc., against their color-equivalents, which he had measured by means of a coarse grating placed in front of the telescope. These diagrams indicated a general tendency for the stars to arrange themselves in a continuous band, with a few separate stars (the giants) forming a small group having red colors and

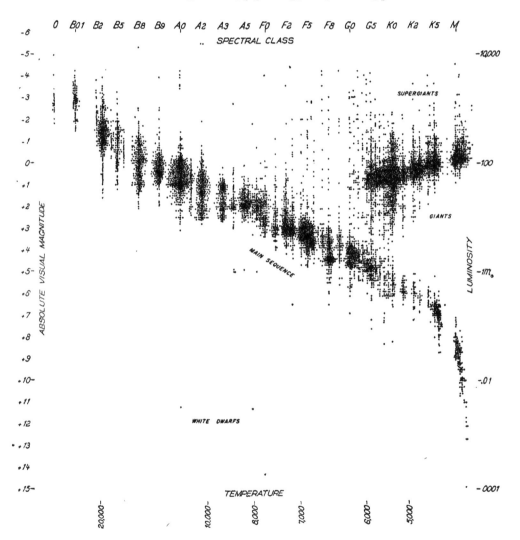

FIGURE 2. The H-R diagram for stars accessible to observation with modern telescopes. This diagram represents a sample of the stars limited only by their apparent magnitude. It does not represent a sample which is complete for a given volume of space. The points represent 6,700 stars. (Diagram by W. Gyllenberg, Lund Observatory, Sweden.)

high luminosities. In 1913, at a meeting of the American Association for the Advancement of Science, in Atlanta, Georgia, Russell for the first time presented a series of diagrams in which a large number of stars with known distances were plotted to show the relation between their luminosities and their spectral types, or temperatures. These diagrams contain all the essential features of similar diagrams made today. They are now usually designated as H-R (or Hertzsprung-Russell) diagrams. In the typical H-R diagram the points occupy only a part of the area. This means that not all pairs of values, L and T, are possible. In fact, Russell pointed out that the diagram looked roughly like an inverted figure 7, though more detailed observations have tended somewhat to remove this similarity. Figure 2 shows this so-called H-R diagram as prepared by W. Gyllenberg of the Lund Observatory in Sweden. We notice that there are two dense sequences of stars: one running from the upper left to the lower right, which is called the main sequence or the sequence of the dwarfs; and another, rather compact, group running almost horizontally near the upper right corner of the diagram. This is the sequence of the giants. In addition, there are a number of stars scattered in the upper portion of the diagram, but only very few are near the bottom. The stars at the top are called supergiants and those at the bottom are the white dwarfs. It will be noticed that the sequence of the giants does not connect with that of the dwarfs. Instead, there is a pronounced gap which has been called the Hertzsprung gap. It looks as though there are no normal giants of spectral types between about A5 and G0. Although there are a few scattered points in the region of the Hertzsprung gap, this is certainly a region of greatly reduced star density.

A diagram of the kind reproduced in Figure 2 is affected by many errors of measurement. For example, the clustering of the stars along several vertical lines is not physically real, but is the result of the grouping of the spectra of the stars into distinct classes. In reality, the spectroscopic criteria change continuously, and the distribution of the two major sequences should thus be quite smooth. There are also great uncertainties in the distances of the stars, which are required in order to convert the observed apparent magnitudes into computed absolute magnitudes. Then there are also uncertainties in the observed apparent magnitudes of the fainter stars, many of which have not been accurately measured but have only been roughly estimated on photographs or by means of visual observations. Finally, it is not always easy to determine the spectral type of a star. For example, on spectrograms of small dispersion it is easy to make a mistake and call an F-type star B0. The reason for this is that in spectral type F the metallic lines are still faint while the H lines are approximately of the same intensity as in class B0. Hence, if the metallic lines are not clearly seen and if the Ca lines near wave lengths λ 3933 and λ 3968 are not in a sufficiently exposed part of the spectrum, a confusion can arise. There are also a number of other considerations which sometimes make it difficult to assign the spectral type to a given star. For example, in some of the hotter stars the spectral lines may be exceedingly weak so that occasionally no lines other than those of H can be seen, even with relatively high dispersion. Finally, there are peculiar stars whose spectral characteristics make it difficult or impossible to find a place for them within the classical Harvard scheme of spectral classification. All these errors combine to increase the spread, in both coordinates, of the principal groups within the H-R diagram. It is necessary to make a searching study to determine accurately the true physical spread of the main sequence, as well as that of the giants.

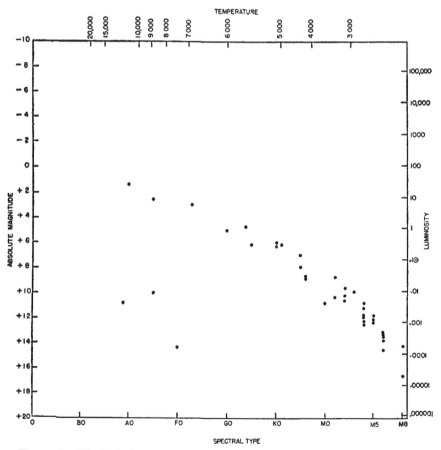

FIGURE 3. The H-R diagram for stars of the solar neighborhood whose distance is less than 5 parsecs.

PLATE II. A. S. Eddington (1882–1944).

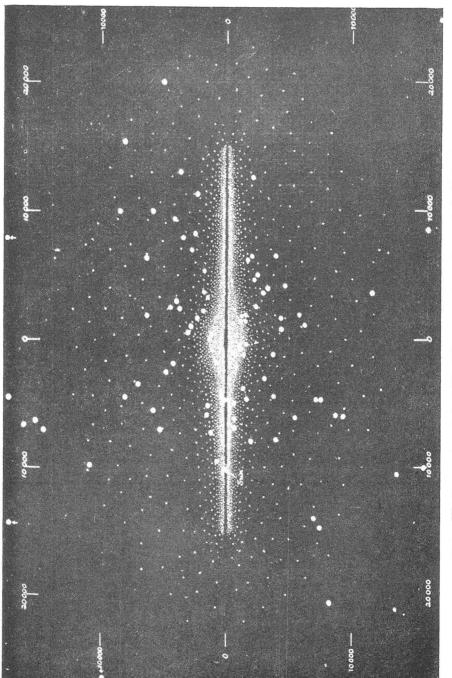

PLATE III. Cross-section of the Milky Way system (according to J. S. Plaskett).

But there is another even more important point which must be remembered in discussing the meaning of this diagram. Gyllenberg plotted the entire material that was available to him on stellar absolute magnitudes and spectral types. This included all stars down to a certain limit of apparent brightness. For example, in the great Harvard Catalogue of Stellar Spectra all stars have been listed whose apparent photographic magnitudes are brighter than about 8.5. In addition, many fainter ones have been plotted, but no special selection was made in picking them out of the mass of faint stars of the same apparent magnitude. Hence, the diagram gives preference to the intrinsically luminous stars which we can see at great distances, and it penalizes the intrinsically faint stars which become invisible at relatively small distances. It is obvious that if we had limited the material, not by the apparent magnitudes of the stars, but by a definite volume of space, we would have observed a greater number of stars at the bottom of the figure and we would have excluded many of the stars near the top. Figure 3 shows a diagram, plotted from a list of stars published by P. van de Kamp in 1940, which are known to lie within a radius of five parsecs from the sun. The total number of these stars is quite small. Only 39 dots occur in the diagram. The main sequence is well represented, and we notice that the concentration is greatest for the intrinsically faintest stars of spectral classes G, K, and M. In Figure 2 the main sequence appeared to thin out precisely at these spectral types. Figure 3 also contains three white dwarfs which are the same stars as those recorded in Figure 2. Hence, we notice that the white dwarfs of Figure 2 were all contained within the small sphere of five parsecs in radius. At distances greater than five parsecs a white dwarf would be so faint that it could not be readily observed with the existing spectrographs. Figure 3 shows no giants and only three stars of spectral classes earlier than G.

Our next diagram, Figure 4, by G. P. Kuiper, contains all the stars located within a sphere around the sun having a radius of ten parsecs. This is twice the radius used in the previous diagram, and the volume included in this study is eight times larger. We see at once that the total number of stars within the diagram has been greatly increased and this time there is one point in the region of the giants. The earliest spectral type is still A0. No B-type stars are included; only four are included which are earlier than F, as against two in Figure 3. The increase in population is particularly noticeable among the cooler stars, that is, those of spectral types K0 and later. There are also more white dwarfs. The spread along the vertical coordinate of the main sequence is very much smaller than that of Figure 2, and this is due to the greater accuracy of the stellar distances and spectral types which were used by Kuiper in the construction of the diagram.

However, it is certain that even this relatively narrow spread is still influenced by appreciable errors in the distances, and the true spread must be still smaller. But Kuiper has reached the conclusion that the spread is not entirely due to errors of observation and that a real physical dispersion exists within the main sequence as it exists among the giant stars. He has noted that several of the best-determined stars lie either slightly above or slightly below the central line of the

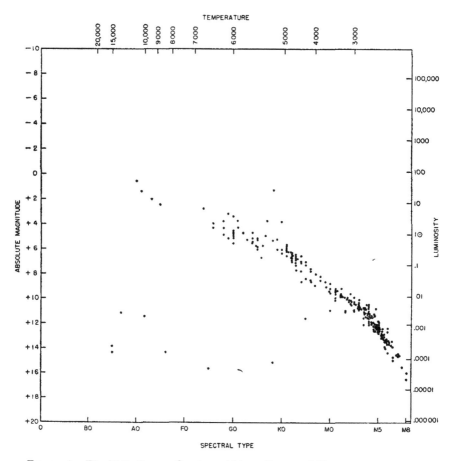

FIGURE 4.   The H-R diagram for stars within a distance of 10 parsecs.

main sequence. A good example of the latter group is Tau Ceti, whose spectral type is G5 and whose absolute magnitude is 6.0. Another example is 61 Cygni, whose parallax was the first accurately determined by the trigonometric method. This is a double star of spectral types K3 and K5 and of absolute magnitudes 7.7 and 8.4. Both components lie slightly below the central line of the main sequence. A star which appears to be noticeably above the main sequence is BD +20°2465, whose spectral type is M5 and whose absolute magnitude is 10.9.

Because of the many uncertainties which enter into the determination of the stellar distances, it is impossible at the present time to determine accurately the amount of the dispersion, but a rough estimate places it at about one stellar magnitude along the vertical coordinate. The spectral types were determined by Kuiper from plate material secured at the McDonald Observatory, and they are the best available at the present time. This classification is based entirely upon the relative intensities of some strong absorption lines which are known empirically to change rapidly in strength along the spectral sequence. The M-type stars were classified by the strength of the red TiO bands, which made it possible to obtain consistent results even for stars which are as faint as the eighteenth photographic magnitude. The reality of the dispersion in absolute magnitude for each spectral class seems to be supported by the discovery that emission lines of Ca II have a tendency to be slightly stronger for those stars which lie above the main sequence than for those which lie below it. There is also a general tendency of the emission lines to increase in strength in the cooler spectral subdivisions.

It is at once noticeable that the dispersion is greatest among the earlier spectral classes. Eggen has found almost no dispersion among the main-sequence stars later in type than about F5, and closer to us than ten parsecs. But for the hotter stars the dispersion is appreciable. Undoubtedly much of this dispersion is real. Two stars in the diagram are so high above the main sequence that they can be described as subgiants. They are Delta Eridani, with spectral type K0 and absolute magnitude 3.8 and Mu Herculis, a triple system whose brightest component is of spectral type G7 and of absolute magnitude 3.7. Three other stars have been designated by Kuiper as "bright dwarfs." They are the double stars Zeta Herculis whose brighter component is of spectral type G0 and of absolute magnitude 3.3. We already commented upon the anomalously high luminosity of this star when we discussed the mass-luminosity relation. Similarly, Eta Bootis, of spectral type F9 and absolute magnitude 3.2, is noticeably above the main sequence, as is also Alpha Canis Minoris, or Procyon, a famous double star whose brighter component is of spectral type F4 and of absolute magnitude 2.8. All these stars have very accurate parallaxes, and their position in the H-R diagram is accurately known. In addition to the narrow main sequence, Kuiper and, a little later, P. P. Parenago, independently discovered the existence of a sequence running approximately parallel to the main sequence and about two magnitudes below it. The members of this sequence are described as subdwarfs and it is probable that Figure 4 contains several representatives of this group. These stars are all characterized by very large space velocities. For many the radial velocities have been measured by means of the Doppler effect, while for some accurate proper motions across the line of sight are also known. It is quite certain that these stars form a group that differs in its kinematic properties from the stars of the main sequence. They are undoubtedly all members of the class of high-velocity stars which are believed to have come into the vicinity of the sun from the central regions of our galactic system. The spectra of the subdwarfs also differ from those of the ordinary main-sequence dwarfs. Those whose types are earlier than about G5 have unusually weak Balmer lines of hydrogen and also weak metallic lines. The vertical dispersion of the subdwarf sequence is much larger than that of the normal main sequence. It is probably larger than two stellar magnitudes. Since the mean line of the subdwarfs runs about two magnitudes below the true main sequence, there are some high-velocity stars which differ little in luminosity from those of the main sequence.

In order to comprehend the significance of these high-velocity subdwarfs we must remember that the great majority of the stars plotted in any of the H-R diagrams we have discussed thus far are located in the vicinity of the solar system and belong to what W. Baade has aptly called "the local swimming hole" of our galaxy. Plate III presents a cross-section of the Milky Way drawn by J. S. Plaskett. The sun is located about 10,000 parsecs from the center, in what we now believe to be a spiral arm similar to arms of other spiral galaxies. The entire system rotates around the center with velocities which are not those of a rigid body, but which follow the laws of Kepler—as though most of the mass of the galaxy, roughly 2×10^{11} solar masses, were concentrated in its central bulge. At the distance of 10,000 parsecs from the center, the velocity of rotation of the system is about 275 km/sec. Most of the stars in the local swimming hole, including the sun, share this velocity; hence, we measure only small departures from the common stream-motion. These stars have approximately circular orbits, and they remain together for a long time, certainly longer than the period of 2×10^{8} years required for their complete revolution around the center. But, in addition to these slow-moving stars (their velocities with respect to the sun are of the order of 20 km/sec) there are some stars whose velocities are as large as 500 km/sec with respect to the sun. The distribution of the motions of the normal slow-moving objects is a random one. But stars whose space velocities exceed 65 km/sec move, with respect to the sun, predominantly towards one hemisphere—that which is opposite to the direction of the circular motion of the local swimming hole. J. H. Oort and, later, G. Miczaika have shown that this tendency is caused by the fact that a star whose velocity exceeds by about 65 km/sec the velocity of 275 km/sec of the local swimming hole, in the direction of Cygnus, at right angles to that of the galactic center in Sagittarius, would escape from the system because such a velocity would be parabolic. The so-called high-velocity stars are therefore in reality slow-moving stars with respect to the center of the galaxy. Their orbits are greatly elongated and they happen to be passing through the local swimming hole, without belonging to it, along paths which favor directions roughly toward the galactic center and away from it, as seen from the solar system. These stars are near their aphelia when they pass through the local swimming hole. Some 10^8 years ago they were near their perihelia, in the vicinity of the galactic center. It is therefore probable that they were formed in the central regions of the Milky Way, and any physical differences between them and the members of the local swimming hole must indicate differences in the composition of the original medium from which the stars were formed.

The stars with roughly circular orbits around the galactic center remain for a much longer time in the vicinity of each other. These stars have a relatively small dispersion in their motions and they form, according to G. Strömberg, the highly flattened system of the Milky Way. The absorbing clouds of interstellar dust and gas are intimately connected with this flattened system. The high-velocity stars, such as the cluster-type variables, show a much larger dispersion in their motions. They are much less condensed towards the central plane of the Milky Way, as are also the long-period M-type variables, and especially the globular clusters. The latter, shown in Plate III by the large white dots, have an almost spherical distribution, and they do not share in the general phenomenon of galactic rotation. In fact, N. U. Mayall, at the Lick Observatory, has measured the galactic rotation of the local swimming hole by referring the measured motions of the globular clusters to the centroid of the local group of stars.

We shall consider next H-R diagrams for several galactic clusters. We have in Figures 5 and 6 two such diagrams prepared from observations at the Göttingen Observatory. The first is for the cluster of the Pleiades, the second for that of Praesepe. In these two diagrams the uncertainty of the vertical coordinate is very small, when we consider the relative positions of the stars in the plot. The distances from us of the stars in each cluster are all very nearly the same, so that their apparent magnitudes furnish us exact differences of absolute magnitudes, even though the numerical value of the distance may not be known with great precision. In other words,

$$\Delta M = \Delta m,$$

a relation which is of great importance in many astronomical investigations. It can be used to advantage not only in the case of clusters, but also for double stars, as has been done, for example, by G. Shajn in Russia and by R. M. Petrie in Canada. The uncertainty of the distance leaves us somewhat in doubt as to the zero-point of the absolute magnitudes. An error in zero-point would shift the entire curve up or down, but would not alter the relative positions of the points in the diagram. The observed quantities which led to the temperatures, or spectral types, were the color-indices of the stars obtained by measuring the difference

$$m_{\text{visual}} - m_{\text{photo}} = \text{Color Index.}$$

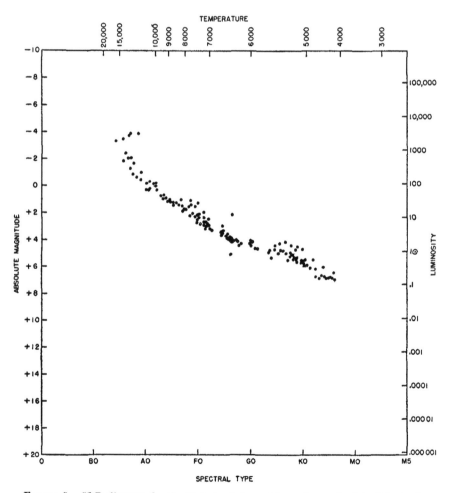

FIGURE 5. H-R diagram for the Pleiades (adapted from the work of A. Behr in Göttingen).

This quantity is a measure of the temperature and has been converted into the latter in the abscissas of Figures 5 and 6.

The resulting diagrams are interesting because their main sequences have an even smaller vertical dispersion than the best H-R diagram obtained for ordinary galactic stars (Figure 4). O. Heckmann and H. Haffner concluded from their discussion of the diagram for the cluster Praesepe that the spread is practically zero, with the exception of a few stars which fall slightly above the limiting line drawn as the lower edge for the main sequence. These anomalous stars above the main sequence have no corresponding objects below it. The skew distribution must be due to the binary nature of these stars, which makes them composite in total brightness as well as in their energy

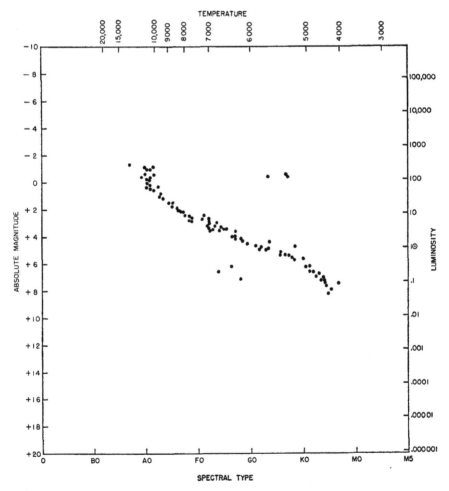

FIGURE 6. H-R diagram for Praesepe (adapted from the work of O. Heckmann and H. Haffner at Göttingen).

distribution. A few stars which are known to be close visual binaries, but which were not resolved in the Göttingen photographs show this effect clearly. Hence, if it were possible to resolve the binaries completely, the dispersion of the main sequence would be even smaller.

The most recent study of this type was made by O. J. Eggen at the University of Wisconsin, using a photoelectric photometer. For the cooler stars of several clusters, including the Hyades in the constellation Taurus, he found the main sequence to have a dispersion of 0.5 magnitude for a given temperature. But the individual distances of the Hyades are not exactly the same, and the small differences of their distances fully account for the dispersion. The lower part of the main sequence of the cluster is actually a line, not a band of finite width as in the galaxy.

But this is true only of the cooler stars in each cluster. We can see that in Figures 5 and 6 the dispersion is appreciably greater among the hottest stars of each cluster. In the case of the Pleiades the brightest stars, namely those brighter than about absolute magnitude 0, scatter with a spread of perhaps two magnitudes. Notice that this occurs for stars whose temperatures are between 10,000° and 15,000°. In the case of Praesepe the scatter is also about two magnitudes, but here it is observed among stars whose absolute magnitudes are brighter than +2 and whose temperatures are close to 10,000°. In the Hyades, Eggen found a similar dispersion for stars brighter than about absolute magnitude +3. These stars seem to belong to two sequences. The main sequence splits into two branches at its upper end; the upper branch consists of relatively luminous A-type stars, the brightest being Theta-two Tauri. This branch contains most of the so-called metallic-line stars, which have peculiar spectra and are characterized by conflicting giant and dwarf features. We must conclude that the brighter stars of each cluster fail to give a unique relation between temperature and luminosity. We shall see that this is precisely what we should expect if the chemical compositions of the stars are not identically the same. In the galaxy, at large, we observe a mixture of stars having the characteristics of the individual clusters, and this accounts for the dispersion of about one magnitude found by Kuiper.

Historically, these differences in the H-R diagrams of galactic clusters were first systematically investigated by Trumpler at the Lick Observatory. We shall return to the discussion of his cluster-diagrams in connection with the question of the hydrogen content of the clusters.

It is important to know the space density of stars within each of the sequences of the H-R diagram. Although actual measurements are available for only the local swimming hole, we can determine, roughly, the density gradient for each group of stars toward the galactic center and away from it; similarly, we can determine the density gradient for different distances above and below the central plane of the galaxy. Then, by extrapolation we can compute the approximate number of stars of each kind. These computations are crude, and do not entirely take account of differences in the populations in the spiral arms and between them, or in the solar neighborhood and in the central bulge. But the results are sufficient to give us a general idea of the frequency of different kinds of stars. By far the greatest number of stars belong to the main sequence. There may be as many as 10^{11} objects of this kind. Next in population, strange as it may seem, is the white-dwarf sequence with a total number which we can estimate at perhaps 10^9 to 10^{10}. The ordinary giants number only approximately 10^7. We do not yet know much about the total number of subdwarfs, but it is quite probable that this sequence is fairly well represented in the galaxy. Their number may be of the order of 10^9 or 10^{10}, but this is uncertain. The supergiants are very rare: their total number may not exceed 10^4. The subgiants are also rare. Along the main sequence the density increases by a factor of the order of 100, as we pass from the top to the bottom.

The enormous preponderance of the stars of the main sequence must have an important bearing upon the question of their origin and evolution. Clearly, the vast majority of stars spend most of their lives within the main sequence. If they wander off at all, they either do so in very small numbers, or they pass only a negligible fraction of their lives in other regions of the H-R diagram.

With the exception of the subdwarfs and high-velocity stars of all kinds, the various groups in the H-R diagrams of Figures 2 to 6 represent the populations belonging to the solar neighborhood. But it has been known for many years that if the stars in a globular star cluster (for example, M 13 in Hercules) are plotted in the same manner, the resulting diagram differs greatly from that which is obtained in the local swimming hole. More recently, Baade at Mount Wilson has shown that the central regions of our galaxy also give a diagram similar to that of the globular clusters. Finally, it is probable that this kind of diagram also applies to the general smooth background of stars which surround the central bulge uniformly in all directions. We designate the stars of the spiral arms, in which our local swimming hole is located, as Population I. The stars belonging to the central bulge and the smooth background form Population II. The two diagrams are superposed in Figure 7. The differences are very marked. Of special interest is the bridge across the Hertzsprung gap of the ordinary H-R diagram. In this band we find the cluster-type variables, a kind of pulsating stars with periods of between 0.1 and

0.6 day which are almost all characterized by very large radial velocities (up to 500 km/sec with respect to the sun) and which derive their name from the fact that they are especially frequent in globular clusters. But they occur also in the galaxy at large, and show all earmarks of Population II. The forking of the giant branch in Population II and the steeper slope upwards and toward the right distinguish these stars from the giants of Population I.

It is sometimes convenient to consider the H-R diagram with regard to the masses and radii of the stars. For this purpose, let us examine a diagram constructed in the usual manner with the ordinate giving the absolute bolometric magnitude or luminosity and the abscissa

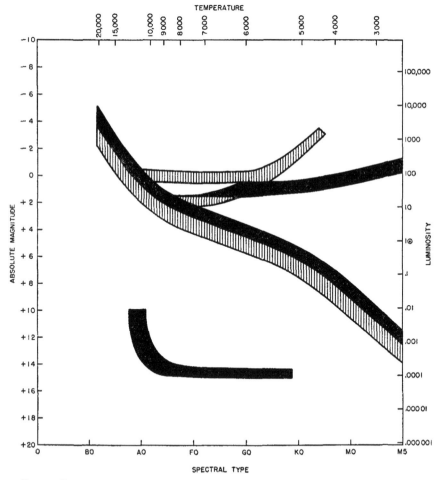

FIGURE 7. H-R diagram for Baade's Population I (solar neighborhood; spiral arms), solid black; and for Population II (central bulge of galaxy, high-velocity stars, globular clusters), shaded.

giving the scale of temperatures. This abscissa can be converted into a set of spectral types ranging from O on the left side to approximately M on the right side. In addition, however, we can show the lines of equal radius and of equal mass. The radii are connected with the bolometric magnitude by means of the expression

$$M_b = +42.63 - 5 \log R - 10 \log T,$$

which is obtained immediately from the formula connecting the luminosity of a star with the temperature

$$L_b = 4\pi R^2 \sigma T^4.$$

The value of the constant in the first expression, namely $+42.63$, is not important for our purpose. The diagram also contains the lines of equal mass. These are obtained from the ordinary mass-luminosity relation, which we consider as an established observational result. If we now enter in this diagram the main sequence of the stars and the giant sequence, we obtain the diagram in Figure 8. The main sequence runs from the top at the left to the bottom at the right and crosses the lines of equal mass, so that at the top the masses are thirty or more times that of the sun, while at the bottom they are only a fraction of that of the sun. The giant sequence runs at first roughly along a line of constant mass ($\mathfrak{M} = 3\mathfrak{M}_\odot$) and then rises toward the right-hand side to larger masses. For both main and giant sequences we note a change in radius; in the former over a range by a factor of 10; in the latter over a range by a factor of 100.

The white dwarfs, also shown in Figure 8, with temperatures between 15,000° and 5,000° and absolute magnitudes between about $+10$ and $+15$, form a wide band with radii of the order of 0.01 of that of the sun. The lines of equal mass do not apply to these stars because they violate the mass-luminosity relation. About 100 white dwarfs are now known, largely through the work of W. J. Luyten at the University of Minnesota. A surprisingly large number of them are members of visual binaries, the other components usually being ordinary red dwarfs. But one of Luyten's double stars consists of two white dwarfs.

It is of interest to speculate upon the existence of close spectroscopic binaries consisting of white dwarfs. We have seen that in Fig. 22 the distribution of the systems in P breaks off rather abruptly, on the left side, near $P = 1$ day. It can be shown that this limiting period corresponds to a system in which two normal stars revolve with their surfaces almost touching. The smallest possible distance between their centers is $R_1 + R_2$, or if the stars are of equal size, $2R_1$. But if one component is normal, while the other is of very great density the limiting distance is but little in excess of R_1. There are indications that such systems may exist and that, among the B-type primaries they could have periods as short as 4 to 6 hours, with mass-ratios of the order of $\alpha = 50$ or even 100. Two white dwarfs could revolve in periods of the order of one minute, or even less, but their detection would have to be made by means of photometric observations. It is possible that eclipsing binaries of this kind really exist. Thus far, the shortest period detected is that of UX Ursae Majoris (see p. 258).

7. The Hydrogen Content of Stellar Interiors

Let us assume that a normal star whose mass is $\mathfrak{M}$ possesses certain internal sources of energy which produce a total luminosity L. The

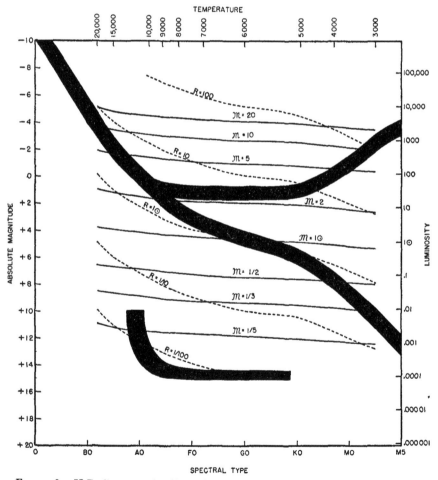

FIGURE 8. H-R diagram with lines of equal mass and equal radius (adapted from A. Unsöld). The white dwarfs depart from the mass-luminosity relation, and the lines of equal mass have no meaning for these stars.

gas pressure, which is proportional to the temperature, combined with the radiation pressure, which also depends upon this temperature, must balance the force of gravitation at all points within the star. Otherwise it would not be in equilibrium. This results in blowing up the star to a certain radius R which is just sufficient to give the required equilibrium. In other words, there must exist a relation

$$\psi(L, \mathfrak{M}, R) = 0.$$

The form of this function has been investigated by A. S. Eddington for stars which are built of ideal gases. It turns out that there exists a simple relation between the absolute bolometric magnitude and the mass, which can be written in the form

$$M_b + 2 \log (T/5200) = f(\log \mathfrak{M}).$$

The factor $2 \log (T/5200)$ represents a small correction which modifies the relation between the bolometric absolute magnitude and the mass of the star. If this correcting factor were absent we would have a one to one relationship between M_b and $\mathfrak{M}$. This would mean that, within the framework of the H-R diagram, lines of equal mass would be represented by horizontal straight lines. However, the correction depending upon T brings about a small slant in the curves so that the lines of equal mass are now represented by a series of curves which are shown in Figure 8. In accordance with the theory of the stellar interior H. N. Russell and H. Vogt found independently that stars which are built on the same model and have the same equation of state as well as the same chemical composition, obey very closely the two relations

$$L = L(\mathfrak{M}); \qquad R = R(\mathfrak{M}).$$

The first relation is identical with the mass-luminosity relation, except for the small factor depending upon T. The second relation can be modified through the introduction of the expression

$$L = 4\pi R^2 \sigma T^4,$$

so that instead of having a mass-radius relation we can eliminate the radius and obtain a luminosity-temperature relation

$$L = L(T).$$

Finally, we already know that the temperature can be represented as a function of the spectral type $T = f(Sp)$, so that the final relation is one between luminosity and spectral type $L = L(Sp)$. This is the main sequence of the H-R diagram. In each of these fundamental relations there enters as a parameter the average atomic weight of the gas and this is dependent upon the abundance of hydrogen. The chemical composition of the stellar material affects the pressure as well as the opacity. Hence, it is important to know the average atomic weight of the substance. If this substance consisted entirely of hydrogen, each atom could be ionized only once and would thus be broken up into two constituents. The average molecular weight of the hydrogen atom is 1, but the average weight per particle of the ionized hydrogen gas would be 1/2. The result would be different if the substance consisted of a heavy element such as Fe, each atom of which has a mass 56 times that of the hydrogen atom. The Fe atom has 26 electrons which it can lose, each having a mass of 1/1845 times the mass of the hydrogen atom. Hence if the Fe gas were completely ionized, each atom would provide 26 electrons and 1 completely stripped nucleus—a total of 27 particles. The average weight of the substance per particle would then be 56/27 or slightly more than twice the mass of the hydrogen atom. Since, as a general rule, the atomic number of each element is approximately equal to 1/2 the atomic weight, the heavy elements would, if completely or nearly completely ionized, all give an average weight of two units. Hydrogen, as we have seen, would give an average weight of 1/2 unit and helium with its two electrons would give an average weight of 4/3. Hence the variation in molecular weight ranges only from 1/2 to about 2, despite the fact that the normal un-ionized atoms have a range of atomic weight from 1 to 238.

In the adjustment which each star must make in order to have its radius consistent with its mass and energy production, the opacity plays an important role. In the interiors of the stars the principal mechanisms producing absorption of radiation are the photoelectric ionization of the bound electrons and scattering by free electrons. The former mechanism is well known and is of greatest importance when the ionization is sufficiently high to produce a large number of free electrons in the medium which can recombine with stripped nuclei, but not so high as to prevent many electrons from sticking to the stripped nuclei. Electron scattering is most pronounced when there are a large number of free electrons. In the theory of Eddington the luminosity of a star is found to depend upon the mass and the radius in the following manner

$$L = \frac{c}{k_0} \, \mathfrak{M}^{5.5} R^{-0.5} \mu^{7.5},$$

where $R^{-0.5}$ stands in place of the same effect we have previously described as a small temperature correction to the mass-luminosity relation. The factor μ is the average molecular weight which enters with the power 7.5 and therefore plays a dominating role despite the fact that μ varies only between 1/2 and 2. We see at once that if μ is small then L is also small, but when μ is large then L also becomes large. This presupposes that the other two variables, $\mathfrak{M}$ and R, remain constant. The constant k_0 is the coefficient of opacity.

In actual practice it is possible to determine the luminosity of a star as a function of X, the abundance of H by mass. If X is zero, the star consists entirely of the heavier elements. If X is 1 the star consists entirely of hydrogen. If X has an intermediate value then 1-X represents the abundance by mass of the heavier elements for which it is customary to adopt the so-called Russell mixture, which is very similar to the chemical composition of the atmosphere of the sun.

If we suppose that we have observed a star whose mass is $\mathfrak{M}$ and whose radius is R, we can make at first an arbitrary assumption concerning the value of X. With this assumption we can compute next the average molecular weight of the substance μ and this, in turn, permits us to compute the absorption coefficient resulting from the photoelectric ionizations. Next, by means of the theoretical mass-luminosity relation

$$L = \frac{c}{k_0} \mathfrak{M}^{5.5} R^{-0.5} \mu^{7.5},$$

we compute the corresponding intrinsic luminosity L. If L agrees with the observed value then we have correctly guessed X. In reality, this will usually not be the case. For example, for the sun we can compute a series of values L corresponding to different values of X. If we plot these values against X we obtain a curve which for $X = 0$ is more than four magnitudes brighter than the true luminosity of the sun. At $X = 33$ per cent, the computed luminosity agrees with the observed luminosity, and at about $X = 80$ per cent, the computed luminosity is about three magnitudes fainter than the observed luminosity. We infer that the true hydrogen abundance is approximately 33 per cent.

In the more recent investigations, attempts have been made not only to compute the hydrogen abundance, but also to introduce as an unknown, Y, the helium abundance; and in the most recent theoretical work a still further unknown, Z, has been introduced to designate the abundances of several elements of intermediate atomic weight; in particular, oxygen and nitrogen. It is not possible to solve for all values X, Y, and Z when we have access only to the observed luminosity, but with the help of a theoretical expression for the energy generation of the sun, M. Schwarzschild was able to obtain another equation which made it possible to solve for X and Y simultaneously. In order to determine the quantity Z, representing the abundance of the oxygen group, it is necessary to have access to still another relation. In the absence of such a relation Mrs. M. H. Harrison was only able to obtain a set of consistent values of H, He, the oxygen group, and the metals. It was then possible to compare these values with sets of abundances determined spectroscopically for the sun's reversing layer. She obtained satisfactory agreement for the following values:

$$\begin{array}{lll}
\text{H}: 70\% & \text{by weight; } 1600 \text{ atoms} \\
\text{He}: 28\% & \text{by weight; } 160 \text{ atoms} \\
\text{O group}: 1.4\% & \text{by weight; } 2 \text{ atoms} \\
\text{metals}: 0.4\% & \text{by weight; } 0.315 \text{ atoms.}
\end{array}$$

The conclusion from the luminosity of the sun is that there is no difference in the absolute abundance of hydrogen and the important H/He ratio as determined from stellar atmospheres and from the theory of the stellar interior. Similar computations were made by G. Keller, who found that the calculated central condensation increases with the assumed abundance of the oxygen group relative to the metals.

The question now arises whether we can extend this type of inves-

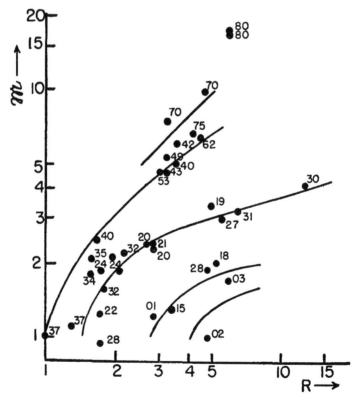

FIGURE 9. B. Strömgren's diagram relating $\mathfrak{M}$ (the ordinate), R, and X (the numbers within the body of the diagram are given in per cent).

tigation to other stars. A first attempt in this direction was made by B. Strömgren who computed, in a manner similar to that employed for the sun, the quantity X for a number of well-observed stars.

In Figure 9 Strömgren has represented the stars in a diagram combining the mass and the radius. For each star the corresponding value of X is written within the body of the diagram. Although the scatter is large, there seems to be a definite tendency for larger values of X to occur at the left and on the top and for small values to concentrate

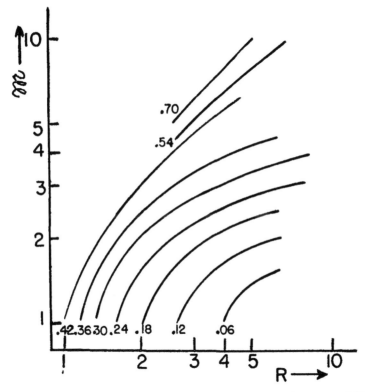

FIGURE 10.   Smoothed curves of equal hydrogen content X in the system of $\mathfrak{M}$ and R coordinates.

near the bottom and at the right. Strömgren next drew in a set of rough curves representing equal values of X (Figure 10). The diagram shows that for stars of the same mass the radius decreases with increasing hydrogen content. Since we already know how to transform the mass and the radius into bolometric absolute magnitude and temperature we can follow Strömgren's procedure and represent the curves of equal hydrogen content within the framework of the conventional H-R diagram. This has been done in Figure 11. It is at once apparent that the curves differ greatly from one another

and there is a striking resemblance between the two first curves at
$X = 0.42$ and $X = 0.36$, and the H-R diagrams of the Pleiades and
Praesepe. This similarity was discussed principally by Kuiper, who
assembled the various H-R diagrams of the galactic clusters, largely
from the work of Trumpler, and presented them in the form of a
combined drawing, a small selection of which appears in Figure 12.
The upper left-hand branches of the curves indeed resemble greatly
the curves derived by Strömgren, but the lower right-hand sections
appear to coincide to a much greater extent than can be reconciled
directly with Strömgren's curves. The latter require a general trend
upward as the H content decreases. In the case of the Hyades, which
according to Kuiper represents a hydrogen-poor cluster, the main
sequence of spectral types F and later is almost exactly in coincidence

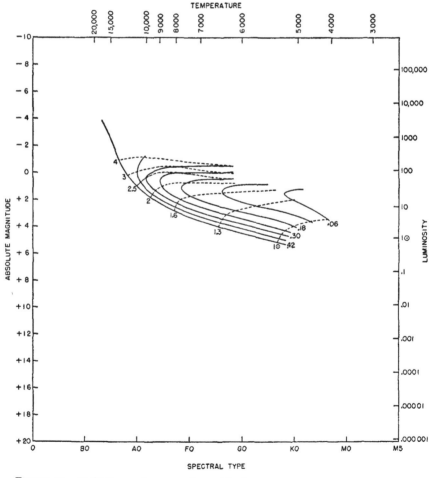

FIGURE 11. B. Strömgren's curves of equal hydrogen content, in the H-R diagram.

with the main sequence found for the galaxy at large. If there is a difference, the Hyades are perhaps a little lower, which is contrary to the trend illustrated in Strömgren's diagram.

As we have already seen, there is a similar departure with hydrogen content from the mass-luminosity relation. For example, the star Zeta Herculis, with $\mathfrak{M} = 0.98\mathfrak{M}_\odot$, $R = 2R_\odot$ and $L = 4L_\odot$ falls above the curve obtained by Kuiper from the data on the masses of the stars and requires $X = 0.11$ or $\mu = 1.45$, while the sun requires $X = 0.35$ or $\mu = 1.00$. We have, thus, two different methods for determining the hydrogen content, provided we are sure that a departure from the mass-luminosity curve or from the main sequence

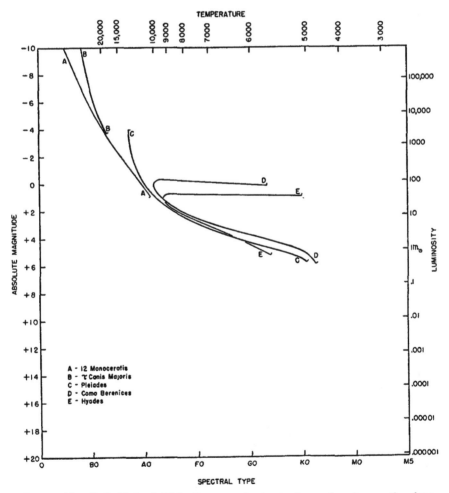

FIGURE 12. G. P. Kuiper's H-R diagrams of galactic clusters based upon the observations of R. J. Trumpler. A, 12 Monocerotis; B, Tau Canis Majoris; C, Pleiades; D, Coma Berenices; E, Hyades.

in the H-R diagram cannot be explained by causes other than chemical composition. In the case of the Hyades, six visual binaries have been discussed by Kuiper in order to provide their masses and luminosities. The weighted mean gives $X = 0.13$, which is regarded as a reliable determination of the hydrogen content of these particular stars, although the scatter of the individual values is quite large: from 0.04 to 0.29. The spectral types of these stars range from A6 to G5 and their visual magnitudes are from 5.7 to 7.1. On the other hand, these same stars when plotted in the H-R diagram fall slightly below the empirical main sequence for the galaxy at large; hence, if we were to interpret their position in terms of hydrogen content we would have to conclude that X is slightly larger than 0.34. This discrepancy has been discussed by Kuiper, who suggested that there may exist, on the main sequence, stars of low hydrogen content in a region which is normally occupied by only stars of much larger mass. He therefore concluded that the divergent portion of the Hyades in the H-R diagram is not a reliable criterion of hydrogen content, while presumably the departure from the mass-luminosity relation could be regarded with considerable confidence.

Since Kuiper's paper was published in 1937, great progress has been made in the interpretation of the stellar energies. We now know that all the stars on the main sequence can be satisfactorily explained in terms of the thermonuclear process of the conversion of hydrogen into helium which was suggested simultaneously by H. Bethe and C. F. von Weizsaecker.

The details of the processes involved are somewhat complicated. A carbon [12] atom captures a proton because of the high kinetic energy of the latter when the temperature is 20,000,000 degrees, and becomes nitrogen [13], which is unstable and disintegrates into carbon [13] and a positron. The atom of carbon [13] captures another proton and becomes nitrogen [14], which in turn captures a proton and becomes the unstable atom oxygen [15]. The latter expels a positron and becomes nitrogen [15]. This also captures a proton which forms one atom of carbon [12] and one of helium [4]. The important conclusion which Bethe derived from the constants of the nuclear processes involved was an expression for the amount of energy produced per unit mass, as a function of the density and the temperature. This provided, in effect, a theoretical expression for L, so that it is no longer necessary to proceed empirically, as Strömgren had done, to determine the relation between $\mathfrak{M}$, R and μ (or X) in order to find L as a function of T and μ only, thus eliminating R. Russell and Moore have made use of the theoretical energy-production in order to interpret the mass-luminosity relation. They found that when μ is constant the agreement is not close. But if μ is assumed to vary smoothly from $0.65(X = 0.68)$ for log $\mathfrak{M} = 1.0(\mathfrak{M} = 10\mathfrak{M}_\odot)$ to $0.97(X = 0.35)$ for log $\mathfrak{M} = -0.5(\mathfrak{M} = 1/3\mathfrak{M}_\odot)$, the agreement is excellent. But it would be premature to conclude that the hot stars of the main sequence are really more abundant in hydrogen than the red dwarfs.

8. The Spectra of Stars in Clusters

Kuiper has attempted to explain the departure of the Hyades from the usual H-R diagram as a deficiency in their hydrogen content. As we have seen, this effect results through the factor μ in the theoretical mass-luminosity relation. The mean molecular weight of the stellar substance enters into the theoretical expression with a power of 7.5. If the star consisted only of hydrogen, its mean molecular weight would be about 0.5. If the stellar material consisted almost entirely of the heavier elements, the mean molecular weight would be approximately 2. Hence, if we have a large amount of hydrogen the luminosity is small. If the hydrogen has been exhausted by the Bethe cycle, then the luminosity must be greater. The time required to convert a major portion of a star's hydrogen supply into helium is quite short—approximately 10^7 years—for the hotter members of the main sequence, namely, those whose spectral types are O and B. But for the cooler stars, like the sun, the interval is of the order of 10^{10} years, or even longer; thus, during the entire lifetime of our galaxy only an insignificant fraction of the original hydrogen content of such a star could have been converted into helium. We must then conclude that the cooler stars, if they differ at all in hydrogen content, must do so as the result of differences in the original composition of the medium from which these stars were formed. On the other hand, for stars of relatively high temperature the nuclear processes are working at such a rapid rate that their present composition is probably largely dependent upon their age and only to a smaller extent upon differences of original composition. Accordingly, Kuiper's interpretation of the Hyades and other galactic clusters would imply a difference in original composition rather than a difference in the ages of these groups.

One way to study this question is to make a detailed comparison of the spectra of stars in different clusters. Such a comparison may or may not yield valuable information concerning the chemical composition of different clusters. We do not yet know whether the outer regions of the stars are at all representative of the inner regions where nuclear processes are at work. The opinions of the theoretical astrophysicists differ radically from one another. Some, like F. Hoyle, are of the opinion that the mixing of the elements is complete because of violent convection currents which must be present in all rotating stars. Other theoreticians, for example P. Ledoux, believe that mixing is hardly perceptible. Still others, like G. Gamow, take a more empirical view and point to the existence of lithium in the sun as an indication of complete separation between the inner and the outer regions of a star. Indeed, if lithium is now present in the reversing layer of the sun and if it is continuously exchanged with the material in the sun's interior, then nuclear processes would convert it rapidly—Gamow says in a few minutes—into other elements, and we should not be able to observe any lithium lines at all. In the absence of any definite information concerning this point we must be prepared for both possibilities.

An attempt has been made to study the spectra of the Pleiades and of the Hyades. The method consisted in comparing as accurately as is possible the spectra of stars at corresponding points of their H-R diagrams. The spectral types of the Pleiades which were available for this test range from B6 to K2. The Hyades begin at Class A1 and extend as far as F9 along the main sequence. In addition, the Hyades contain four well-known red giants, whose spectral types are between G7 and K0. There are no late-type giants in the Pleiades and there are no B-type stars among the Hyades.

If we compare critically the main sequences of the two clusters, we find that at F9 and F8 the spectra are almost identically the same; even a somewhat superficial study fails to reveal systematic differences between the two groups. At the other end of the main sequence, where the two clusters overlap, namely at A1, the difference is most conspicuous and is definitely systematic in character. The Hyades are all much richer in the metallic lines, including not only Fe I and Fe II, but also Ca I, Sr II, and probably Mg II and Si II. At the same time, the lines of H and probably those of Ca II are much weaker in the Hyades than in the Pleiades. This effect is somewhat obscured in the case of the Pleiades, because of the large rotational velocities which are frequently found in that cluster, but even allowing for the rotational broadening, it is a most conspicuous phenomenon and one that cannot be attributed to observational effects. It is closely connected with the well-known tendency of the Hyades to possess members which W. W. Morgan described as having metallic-line spectra. It is, of course, also known that these metallic-line spectra are not present in the Pleiades.

We have the impression that although not all the early and middle A's among the Hyades have been recognized as metallic-line stars, there seems to be a general tendency of most members in this direction. It is true that there exists among the Hyades a marked dispersion in their characteristics, ranging between a behavior which differs only slightly from that of the Pleiades to that of the typical metallic-line objects. The latter have all been classified by Morgan as A stars and not as F stars. If, following J. L. Greenstein, we had described these stars as members of class F, we should have found them to differ materially from the corresponding F stars of the Pleiades, as well as from those normal F stars which Morgan has found in the Hyades. It is not necessary for us to decide at this stage which classification is the more correct one. The important thing is to recognize that there exists a striking systematic difference between the spectra of the Hyades and those of the Pleiades, and that this difference is pronounced among the earlier types and disappears in the later ones.

It is of interest in this connection that some recent observers believe that the main sequence of the Hyades falls slightly below that of the Pleiades and even below that of the galactic stars in general. This tendency is most conspicuously shown in the work of J. M. Ramberg, who has found that the departure is greatest among the A's, F's, and G's, but disappears not far from K0. Although the systematic difference in the absorption-line spectra to which we have referred is hardly noticeable at F8 and even at F5, it may be that the two phenomena are related. However, R. E. Wilson believes that the main sequences of both clusters agree with that of the galaxy, except at the upper ends, and this is supported by Eggen, who finds that the main sequence of the Hyades coincides with that of the stars closer than ten parsecs, for spectra later than about F5 or G0.

The difficulty of classifying the metallic-line stars illustrates the problem we face when we find conflicting spectroscopic criteria in stars of different hydrogen content. It is difficult and perhaps impossible to decide whether the tendency to develop the characteristic features of the metallic-line spectra may in some way be related to the low hydrogen abundance which has been suggested for the Hyades. The criteria, however, are not unreasonable if considered from this point of view. Rather than attempt to predict theoretically what a star's spectrum would look like if the hydrogen abundance were small, we might consider the observed data in the light of information gathered from such a star as Upsilon Sagittarii. Greenstein's work has shown that if we analyze the spectrum by the conventional methods we find a small hydrogen abundance and a relatively high abundance of helium and of the Russell mixture of heavy elements. Qualitatively, this is the sort of thing we have observed in the metallic-line stars. They imitate stars of higher luminosity because they have relatively weak hydrogen lines, strong lines of other elements, spectroscopic characteristics of higher than average luminosity, and appreciable turbulence. It is not surprising that Greenstein's spectrophotometric study of the metallic-line star Tau Ursae Majoris brings out precisely these characteristics. A difficulty is encountered when we consider the departures of these stars from the H-R diagram. Are they members of the main sequence, as previous workers have assumed, or are they several magnitudes brighter? Strömgren's original study of hydrogen abundances led us to expect that the main sequence of the hydrogen-poor clusters would lie higher than the main sequence of a hydrogen-rich cluster, but Kuiper's investigation in 1937 and Ramberg's, more recently, have placed the main sequence of the Hyades lower than that of other stars. It seems that there is a serious difficulty because of our lack of knowledge of the precise relation between spectrum and temperature. Strömgren's curves related L with T. It is obviously not safe to apply to a hydrogen-poor cluster the same relation between T and Sp that was found correct for hydrogen-rich stars. This difficulty is illustrated by the uncertainty of placing the metallic-line stars in the H-R diagram. If they are A's, then they cause the main sequence of the Hyades to curve downward at the earliest types, as it does in the work of J. Titus and W. W. Morgan. But if we should move these same stars over to the right until they coincide with the middle F's then their absolute magnitudes would suggest an upward curvature of the main sequence, like that found by Eggen. But the abscissa of the H-R diagram is really the temperature. Hence, Eggen's color-measurements should place these stars correctly.

Another difficulty of the interpretation in terms of hydrogen content arises from the great similarity of the spectra of the later subdivisions of class F, the G's and the K's. If we had observed these same spectra without knowing that they came from different clusters, we should certainly have concluded that they belong to the same spectral type and luminosity. We would then have obtained identical compositions from the equivalent widths of the spectral lines. The similarity of these spectra is really quite striking. We must distinguish between the problem of determining absolute abundances and differences in abundance. The former is difficult and can at the very best be carried out with uncertainties of the order of a factor of two. But the latter is much easier. If the curves of growth are similar, as they are in these dwarfs, it should be possible to detect differences in abundance of the order of ten per cent. No accurate measures are available, but even a simple comparison of the spectra should readily reveal differences of the order of fifty per cent.

It would be an unexpected and improbable coincidence if in these spectral classes a change in the hydrogen content would so completely simulate the spectral type and the luminosity of a star in a different part of the main sequence in the H-R diagram. Hence, it is difficult to attribute different compositions to the fainter members of the two clusters. But it is probable that such a difference does play a role among the earlier spectral types.

In this connection it may be well to recall again that the B-type members of the Pleiades are systematically quite different from the ordinary galactic B stars. They are certainly more luminous than the main sequence, perhaps by as much as 2 or 3 magnitudes. This fact led Trumpler, and later Kuiper, to draw in the H-R diagram a curve steeply inclined upward on the side of higher temperatures than those corresponding to class A0. The similarity of this H-R diagram with one of Strömgren's curves led Kuiper to the conclusion that the Pleiades had a smaller hydrogen content than the average B star in the Milky Way. It is again very difficult to decide whether the pronounced spectroscopic features of the brighter Pleiades necessarily indicate low hydrogen content in their atmospheres. The spectrum of Maia is particularly interesting because it is the only star of type B in the Pleiades which has practically no rotation. The spectral lines give conflicting results and do not permit us to assign to this star a definite spectral type and luminosity, unless at the same time we recognize the existence of a parameter other than temperature and pressure. On low-dispersion plates, Morgan finds that the hydrogen lines of all the brighter Pleiades are fainter than they would be if the spectral type were assigned in accordance with the ratio He I to Mg II. Since this weakness of the hydrogen lines is consistent with higher luminosity, he is inclined to attribute slightly later types to these brighter Pleiades than was customary when the hydrogen lines themselves served as a criterion of spectral type. This places Maia and the other bright Pleiades among the stars of types B6 to B8 with luminosities of the order of -3 to -2. There are, however, a few other B8 and B9 stars in the Pleiades which do not have these high absolute magnitudes and whose hydrogen lines are normal.

The weakness of the hydrogen lines in Maia, together with the relatively great strength of Mn II and Fe II and the simultaneous strength of He I, all combine to indicate fairly high luminosity, of the order of -2 or -2.5. Again, at least qualitatively, the spectroscopic criteria are in accordance with the hypothesis that the hydrogen content of the atmosphere is relatively low.

Since it must be even lower in the Hyades, it is reasonable to suppose that because of the short lifetimes of the B-type stars they have already exhausted their supply of energy provided by the Bethe cycle and have been transformed into other types of stars. This would account for the absence of B stars in the Hyades and other hydrogen-poor clusters.

Again, as in the case of the Hyades, we must conclude that the spectra of the fainter and cooler Pleiades are not appreciably different from those of the great mass of galactic stars.

If present theories may be relied upon, it would perhaps be reasonable to suppose, as H. Vogt has done in his recent book, that the original content of a cluster was the same for all its members. In the Hyades the B-type stars may have completely exhausted their supply of hydrogen and they have disappeared as B stars. The A's and perhaps early F's are deficient in hydrogen, but the later-type members, whose evolution proceeds much more slowly, have not been greatly affected throughout the history of the cluster. In the Pleiades we must then assume that the B-type stars are in the process of burning up their supply of hydrogen, while the A's, F's, G's, etc., still have their original composition. Finally, the great mass of galactic B and O stars are rich in hydrogen because they are relatively young and we observe them only when they are producing the required amount of energy to give us a spectrum of class B.

In one way this picture differs from that proposed by earlier workers. It assumes that the hydrogen content was at first essentially the same in all clusters and in the great majority of the stars of our galaxy. This is much more satisfactory since nearly all workers agree that the stars are formed out of interstellar matter, and that in the course of their lives they throw back a part of their substance into the diffuse medium. The composition of the medium and that of young stars is about the same. We are probably concerned more with the evolutionary ages of the clusters, which may not necessarily be identical with the ages measured on the ordinary time scale. On this picture then, the Hyades would represent an old cluster, the Pleiades a younger cluster and the double cluster h and Chi Persei, with its great mass of O and B stars, a group of particularly young objects.

The data at our disposal are insufficient to resolve the many difficulties which this or any other hypothesis would involve. A disturbing thought arises immediately when we consider the wide range in L for a given Sp, among the late B stars of the Pleiades and among the A stars of the Hyades, or when we examine the two distinct branches in Eggen's recent H-R diagrams of the Hyades, Pleiades, and Coma clusters. It is not at present possible to overcome these difficulties, except by taking refuge in the fact that conflicting criteria prevent us from assigning these stars to their proper temperature classes. Recent investigations by Eggen of the colors of stars and by Weaver of their spectra have stressed the probable existence of several discrete and intrinsically very narrow sequences in the H-R diagram. It is theoretically not unreasonable that such sequences may exist because stars built on slightly different models may require different intervals of time to convert a given amount of hydrogen into helium. But at the time of this writing the observational evidence has not been fully published, and it has not been sufficiently scrutinized by all astronomers. One of the most fruitful types of investigation would undoubtedly be the extension of Trumpler's work with the help of spectrograms of high dispersion. This can now be done for at least some of the clusters.

This interpretation differs from the one proposed by Kuiper, and it rests largely upon the assumption that identity of spectroscopic features in corresponding points of the H-R diagrams implies identity of chemical composition. We have already seen that there is some uncertainty about the extent of the mixing of the stellar material, but even if there were no such mixing a difference in hydrogen content should produce absolute-magnitude effects due to the greater luminosity of a star whose X is small. Apparently the spectroscopic result is simply another way of restating Kuiper's conclusion that the main sequence of the cooler members of the Hyades almost coincides with the average main sequence of the galaxy. Why it is that the six binaries investigated by Kuiper in the Hyades lie above his empirical mass-luminosity relation cannot now be answered. But we must remember that, at least among the spectroscopic binaries, large departures from the mass-luminosity curve which are not caused by differences in chemical composition have been observed.

Another difficulty arises when we compare the spectrum of Maia with that of Upsilon Sagittarii. The latter is hydrogen-poor in its reversing layer. The former is suspected of having partly exhausted its internal supply of hydrogen by the Bethe process. Both are late B or early A stars, and their effective temperatures are not very different. Yet the spectra are certainly very different. Maia is less luminous than Upsilon Sagittarii, and this may account for part of the difference. But it is hard to see why in Maia all lines should be weaker than usual, while in Upsilon Sagittarii the lines of the ionized metals and of the light gases (He I, etc.) are very greatly strengthened. If in both stars H is deficient, then the continuous opacity must come from He I, the metals, and the free electrons. The general weakening of the lines in Maia suggests that scattering by free electrons—abundant because of high ionization of the metals—may play a role. But that would imply a deficiency also of helium, which is improbable. It is necessary to leave this question unanswered. But the difference in the behavior of the two stars suggests that the causes of their peculiarities are not the same. Perhaps Upsilon Sagittarii has a real deficiency of hydrogen in its atmosphere, while in Maia incomplete mixing produces only an indirect and as yet obscure spectroscopic difference from the ordinary main-sequence stars of the same temperature.

9. The Composition of the Sun's Atmosphere

We turn to the determination of the chemical composition of the outer layers of different stars, remembering that the results may not give us any information concerning the nuclear processes which operate in the stellar interiors. Because of this uncertainty it will be useful to study at the same time the physical properties of the atmospheres of the stars in order to ascertain whether there are present luminosity effects of the kind that may be caused by the exhaustion of hydrogen in the stellar interiors.

In 1914, H. N. Russell noticed a conspicuous parallelism in the relative frequencies of the atoms of the more abundant elements in the stars and the crust of the earth. But real progress in this field had to await the development of the theory of ionization by M. N. Saha and by R. H. Fowler and E. A. Milne. In 1925, Miss Cecilia H. Payne (now Mrs. S. Gaposchkin) applied this theory to the problem of atomic abundances, with the assumption that the number of atoms required to make a spectral line barely visible in a stellar spectrum is the same for all lines of all elements. This method of marginal appearances of the spectral lines consisted in determining at which spectral type, or temperature, a line first becomes visible along the spectral sequence, and at which one it is last seen. Considering the complete lack of information regarding the individual absorbing properties of the various atoms, it is surprising how much could be learned in this manner. The preponderance of the lighter elements was found to be the most striking aspect of Miss Payne's results. The first modern determination of the abundances of the chemical elements in the atmosphere of the sun was made by H. N. Russell (Plate IV) in 1929. The first step in this investigation was to calibrate Rowland's intensity estimates in terms of numbers of atoms. This calibration was accomplished with the help of the theoretical intensities in multiplets. A number of strong lines which had been accurately measured with the microphotometer and for which theoretical transition probabilities were known served for finding the absolute values of the numbers of atoms. In this manner it was found that it requires 6×10^{12} atoms per square centimeter to produce a line of Rowland intensity zero. Of course this is a very crude way of describing the absorbing effects produced by the various atoms, but it sufficed as a first approximation. Having established a satisfactory calibration, Russell next determined a quantity M which represents the "absorbing power of each spectral state" and may be expressed by the following relation:

$$\log M = \log A_0 + \log W - \frac{5040}{T}E.$$

Here A_0 is a constant depending upon the abundance of the element, W is the total statistical weight of the spectroscopic state, and E is the excitation potential. A similar expression can be written for ionized atoms and the two can be brought into agreement by applying the ordinary ionization equation. Russell next obtained the sums of all the values of M for the different energy levels and designated them by S_0 for the neutral atoms and S_1 for the ionized atoms. He then listed the total numbers of atoms of each element in both stages of ionization. These quantities are expressed by the letter T. Ionizations higher than the first could be neglected in the case of the sun. Finally, the total mass of the atoms or molecules of the substance, per unit area of the sun's surface, resulted from the multiplication of T by the atomic weight or by the molecular weight. In several special cases the quantities A_0 and A_1 had to be corrected because the most important spectroscopic levels were out of reach, in the ultraviolet region. For example, in the case of Cd only a single intersystem combination ($^1S_0 - {}^3P_1$) is observed in the sun at λ 3261. For this line the physicists have found that the number of effective resonators is 1/600 of that which would apply to the resonance line ($^1S - {}^1P$). Hence, a correction in the logarithm of $+2.8$ was adopted for Cd. Russell's final results gave the relative abundances of 56 elements and 6 compounds. The entire number of all the metallic atoms above one square centimeter of the sun's photosphere is 8×10^{20}. Their mean atomic weight is 32 and their total mass is 42 mg. The well-known difference discovered by W. D. Harkins in the abundances of even and odd atomic numbers is conspicuous, the former being about ten times more abundant than the latter. The abundance of the non-metals was determined with a comparatively small degree of precision, but Russell was able to establish the very great abundance of hydrogen, which, in the solar atmosphere, constitutes sixty parts by volume, while helium accounts for two parts, oxygen also for two, and the metals only for one. Finally, the free electrons account for only 0.8 part by volume. The extreme rarity of Li and Be is conspicuous. The metals of the second short and first long periods in the table of the elements are far more abundant than the heavier metals. Of interest is the great abundance of Fe and the relatively small abundance of Sc. The logarithmic abundances of the elements are shown in Figure 13, which gives them by mass/sq. cm. of the surface of the sun.

During the past twenty years a number of other investigations of the abundances of the chemical elements in the sun have been published. The most recent is by A. Unsöld, who not only made use of the best theoretical and experimental values of the oscillator strengths but also discussed the effects of turbulence, damping, and variation in continuous absorption with wave length. Unsöld used the measures of equivalent widths made by C. W. Allen and determined for each quantum state s and ionization stage r the total number of atoms, log $N_{r,s}H$, where H is the thickness of the absorbing layer. By means of a

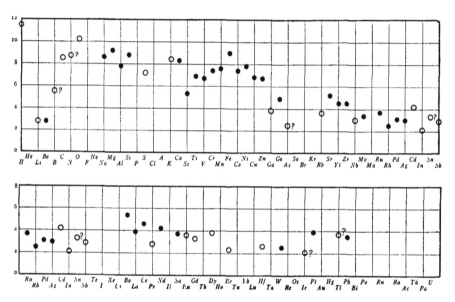

FIGURE 13. The abundances (in logarithms per unit area) of the chemical elements in the atmosphere of the sun (according to H. N. Russell).

comparison of values for different ionization and excitation potentials, he determined next the average level of ionization in the solar reversing layer,

$$I = 8.33 \text{ volts},$$

and the excitation temperature $T = 5675°$. A column having a square centimeter cross section contains a total number of atoms equal to $10^{24.5}$. Within this same column the number of free electrons is $10^{20.7}$, and the mass of the column is 7.4 gms. The final results on the abundances in the sun are shown in Figure 14, while Table II lists the cosmic abundances of the elements in the sun according to investigations by Unsöld, Strömgren, and Russell, as well as in the B-type star Tau Scorpii by Unsöld, and in the meteorites by V. M. Goldschmidt and by I. and W. Noddack. Russell's results in Figure

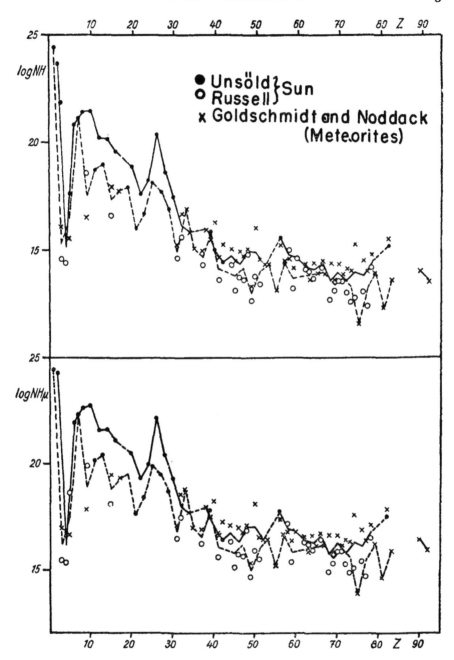

FIGURE 14. The abundances of the chemical elements in the atmosphere of the sun (according to A. Unsöld). The lower diagram represents the atomic abundances multiplied by the atomic masses.

TABLE II

Cosmic Abundances of Elements (log NH)

Z	El.	Sun (Unsöld)	Sun (Strömgren)	Tau Scorpii (Unsöld)	Sun (Russell)	Meteorites (Goldschmidt)	(Noddack)
1	H	24.13	24.65	24.41	23.1		
2	He			23.66	21.6?		
3	Li				14.6:	16.20	16.03
4	Be				14.4	15.50	16.22
5	B				17.6:	15.58	
6	C	20.94		20.65	20.1	17.72	17.88
7	N	21.26		20.99	20.6:		15.88
8	O	21.38		21.40	21.6	20.74	20.60
9	F				18.6:	16.52	
10	Ne			21.46			
11	Na	18.93	18.59		19.8	18.85	18.68
12	Mg	20.16	20.23	20.17	19.9	20.14	20.00
13	Al	18.98		18.97	19.0	19.14	18.98
14	Si	19.94		20.21	20.1	20.20	20.08
15	P				16.6:	17.96	17.94
16	S	19.57			18.3:	19.26	19.20
17	Cl					17.9?	17.60
18	Ar						
19	K	17.85	17.95		19.4:	18.04	18.01
20	Ca	18.88	18.86		19.3	18.96	18.88
21	Sc	15.98			16.2	15.38	16.58
22	Ti	17.61			17.8	17.87	17.83
23	V	16.70			17.6	16.31	16.96
24	Cr	18.23			18.3	18.25	18.20
25	Mn	18.11			18.5	18.02	17.80
26	Fe	20.37			19.8	20.15	20.37
27	Co	17.68			18.2	17.74	18.03
28	Ni	18.60			18.6	18.86	19.22
29	Cu	16.88			17.6	16.86	17.19
30	Zn	17.43			17.5	16.76	16.71
31	Ga				14.6:	15.12	15.30
32	Ge				15.6	16.47	16.83
33	As					Trace	16.88
34	Se					15.38	16.22
35	Br					15.83	14.30
36	Kr						
37	Rb				14.3:	15.03	14.90
38	Sr	16.00			15.9	15.80	16.10
39	Y	15.86			15.2	15.19	15.78
40	Zr	15.02			15.1	16.34	16.22
41	Nb				13.6:		14.64
42	Mo	14.43			14.0	15.18	15.38
43	—						
44	Ru				14.3	14.76	15.38
45	Rh				13.1	14.31	14.71
46	Pd				13.7	14.60	15.26
47	Ag				13.6	14.71	14.78
48	Cd				14.8:	Trace	15.06
49	In				12.6:	13.56	13.03

PLATE IV. Henry Norris Russell of Princeton University.

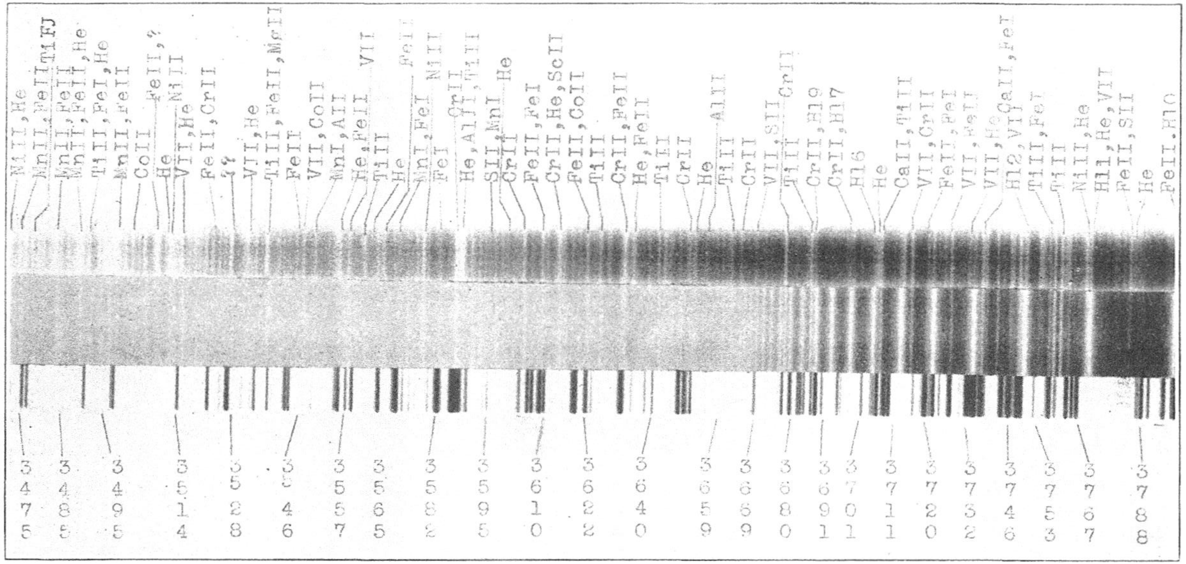

PLATE V. The Spectra of Upsilon Sagittarii and Alpha Cygni. Notice the weakness of the H absorption lines in the former, and also the complete absence of the Balmer discontinuity at λ 3647.

TABLE II (*Continued*)

Z	El.	Sun (Unsöld)	Sun (Strömgren)	Tau Scorpii (Unsöld)	Sun (Russell)	Meteorites (Goldschmidt)	(Noddack)
50	Sn				13.8?	15.66	16.46
51	Sb				13.4:	Trace	14.46
52	Te					?	14.30
53	J					14.33	
54	X						
55	Cs				?	13.20	13.06
56	Ba	15.60			15.9	15.12	15.35
57	La				14.4	14.52	
58	Ce				15.0	14.57?	14.64
59	Pr				13.2:	14.18	
60	Nd				14.6	14.72	14.50
61	Il						
62	Sm				14.1	14.26	14.49
63	Eu				14.0:	13.65	
64	Gd				13.7:	14.42	
65	Tb					13.92	
66	Dy				14.2:	14.51	
67	Ho					13.96	
68	Er				12.7:	14.41	
69	Tu				13.1:	13.66	
70	Yb				13.6:	14.38	
71	Cp				13.6:	13.88	
72	Hf				13.0	14.38	13.92
73	Ta				12.6:		14.08
74	W				12.8	15.36	15.27
75	Re					11.46	11.71
76	Os				13.1:	14.44	14.75
77	Ir				12.4?	13.96	14.10
78	Pt				14.2	14.66	14.96
79	Au					13.96	13.88
80	Hg						
81	Tl					Trace	12.35
82	Pb	15.2			13.8	15.16	15.92
83	Bi					Trace	13.60
90	Th					13.97	14.12
92	U					13.56	

14 are systematically lower than those of Unsöld because the former had used in 1929 Unsöld's classical damping factors instead of values which are about ten times larger. In Table II a correction has been applied for this difference.

10. *The Composition of the Atmospheres of Normal Stars*

When this method was applied to the spectra of several bright stars a strange result was obtained. It was found that when the total absorptions of the spectral lines belonging to a single multiplet were plotted against the theoretical intensities obtained from the sum rules or against the corresponding relative numbers of atoms, the observations defined sections of approximately straight lines of different slopes in different stars. For example, a multiplet of Fe I was found to have a slope of nearly 45° in Alpha Persei, but only of about 20° in Procyon. When the numbers of atoms were plotted logarithmically and the intensities of the absorption lines were expressed in terms of the total energy lost from the continuous spectrum, this phenomenon, which was at first described as "the gradient effect," made it abundantly clear that for accurate work it was not sufficient to rely upon the conversion of Rowland's estimated solar line-intensities into corresponding numbers of absorbing atoms. Instead, a completely new method of conversion had to be found which, for each individual star, would permit us to express the observed intensities of the absorption lines in terms of numbers of atoms.

We now know that the gradient effect is a manifestation of an important physical phenomenon in the atmospheres of the stars which may be described as turbulence. Fairly large volumes of gas move in different directions, and this motion is superposed over the irregular motion of the individual atoms, which in turn depends upon the temperature. We have no definite information concerning the distribution of these turbulent motions. In the absence of a better procedure, it seemed plausible to assume that the motions are distributed at random and follow Maxwell's law. In that case the absorption coefficient within a line has the form corresponding to the Maxwell distribution, characterized by a bell-shaped curve which is flat at the top and has fairly steep sides and relatively weak wings. An assembly of atoms which absorbs the radiation from underlying layers in accordance with this law produces a line which increases in total absorption with the number of atoms within the layer. When there are very few atoms the total absorption is directly proportional to N, but as the spectral line becomes more and more saturated in the center—as must always happen when the number of absorbing atoms increases—the total absorption increases at a less rapid rate. Hence, if we plot the total absorption against the number of absorbing atoms, we obtain a curve which for small values of N approximates a straight line with a slope of 45° and then gradually flattens out, never really becoming horizontal, but having tangents which become more and more nearly horizontal.

In reality the stellar absorption coefficient is never completely determined by the effect of turbulence. There is always present an inherent tendency of all atoms to produce absorption wings whose intensity is inversely proportional to the quantity $(\lambda - \lambda_0)^2$. This broadening is caused by the properties of the absorbing atom. As the atom radiates it loses energy; in terms of the classical description of the process of radiation the amplitude of the wave train of radiation emitted by the atom decreases and gradually dies down, as in a damped oscillator.

We know from the classical theory of the electron that a consistent description of the resulting radiation is obtained if we interpret the process in terms of infinite wave trains of different wave lengths, λ. The damped wave train which corresponds to the radiating atom can be described by means of a Fourier integral containing an infinite variety of wave trains having wave lengths which continuously change from zero to infinity. However, the amplitude of the central wave train, whose wave length is λ_0, is large, while the amplitudes of all other wave trains become rapidly smaller as λ differs more and more from λ_0. This type of broadening is called radiation damping. It can be modified by collisions among the radiating atoms which limit their damped wave trains and cut them off completely if the collisions are real, or change them in phase if the collisions are in the nature of close passages. These mutilated wave trains can also be described in terms of a Fourier integral of a type similar to the Fourier integral in the damped oscillation. The difference consists in the numerical value of the damping constant, which determines the strength of the damping wings in the absorption coefficient.

It so happens that at great distances from the central wave length, λ_0, the wings of the damping part of the absorption coefficient are larger than those produced by the irregular Maxwellian motions, whose form may be described by the expression $e^{-(\lambda - \lambda_0)^2}$. Hence, if we examine the curve which connects the total absorption of a line and the number of atoms, we find that, after it has flattened out to a considerable extent because of the saturation of the Doppler cores of the lines, there is a further rise in the curve as the wings from the damping part become conspicuous. Finally, for very large values of N the curve again approaches a straight line of a smaller slope, and at the limit the total absorption becomes proportional to the $\sqrt{N}$. The complete curve is called the "curve of growth."

For different stars the velocities of the turbulent elements may be quite different. In the sun they are very small. In ordinary main-sequence stars turbulence is never conspicuous, and in most of them it is not detectable with present methods. But in giants and supergiants and especially in some peculiar stars, the turbulent velocities may be very large. They are in fact very much larger than the ordinary thermal motions of the separate atoms. Thus, in the sun and in other main-sequence stars the thermal velocities of atoms of average atomic weight are of the order of 2 km/sec, but in Alpha Persei the turbulent velocities are of the order of 7 or 8 km/sec and in 17 Leporis they are approximately 67 km/sec. In practice we can determine the turbulent velocity if we draw the curve of growth by combining a large number of small sections obtained either from individual multiplets or from super-multiplets in all of which the relative numbers of atoms are known from the theory of the spectra. There is an element of uncertainty in combining different sections of the curve which correspond to different groups of lines. This uncertainty can sometimes be reduced by means of laboratory determinations of the transition probabilities or by means of wave-mechanical computations. In modern spectroscopic work it has become more and more convenient to use R. B. King's laboratory determinations of the transition probabilities, which are now available for several of the more important atoms, and to check the resulting curves of growth by means of transition probabilities which have been inferred from the curve of growth of the sun.

Whichever method is used, the resulting curves for different stars coincide in their lower left-hand parts, where a straight line is approached whose theoretical expression is given by

$$A = 1.8 \times 10^{-13} N.$$

A designates the total absorption of a spectral absorption line in Angstrom units expressed as the equivalent width of a line of zero intensity in the center and with vertical sides, and where N is the number of atoms. The flat part of the curve of growth depends upon the turbulent velocity. If we designate

$$b = \lambda_0 \frac{v_0}{c},$$

where v_0 is the most probable velocity of a turbulent element and c is the velocity of light, then it can be shown that the absorption coefficient of the intermediate (Doppler) portion is

$$\sigma = \frac{\sqrt{\pi} \epsilon^2 \lambda_0^2 N f}{mc^2 b} e^{-\left(\frac{\Delta\lambda}{b}\right)^2},$$

where ϵ is the charge of the electron, m is its mass, N is the number of atoms per cubic centimeter and f is the oscillator strength which measures the transition probability. If turbulence is absent and only thermal motions are present, then we must replace b by

$$b_0 = \frac{\lambda_0}{c} \sqrt{\frac{2RT}{\mu}},$$

where μ is the atomic weight and R is the Boltzmann constant. In the case of pure exponential absorption, the intensity of the emerging radiation is

$$I = I_0 e^{-\sigma H},$$

where H is the thickness of the atmosphere. If the absorption is not exponential, that is, if the re-emitted radiation cannot be neglected, this expression must be changed; but in principle no new factor is introduced. We have merely to replace $I = I_0 e^{-\sigma H}$ by some such expression as

$$I = \frac{I_0}{1 + \sigma H}.$$

The quantity which we actually measure is the equivalent width

$$A = \int_{-\infty}^{+\infty} (1 - I) d\lambda.$$

The integration may be carried out by means of suitable approximations so that in each case a curve of growth can be computed which

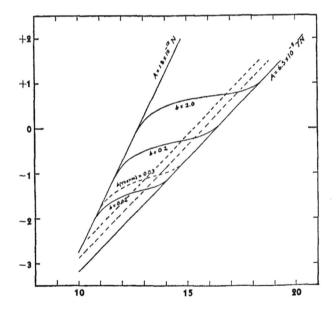

FIGURE 15. Theoretical curves of growth computed for the simplest possible case of exponential absorption (no re-emission) and for three different values of the turbulent velocity. The limiting straight line on the right corresponds to the classical value of the damping constant. The dotted straight lines are for two larger values of the damping constant: four and ten times the classical value. The dotted curve for $b = 0.03$ corresponds approximately to the thermal curve of growth when $T = 10,000°K$. The ordinates are values of log A, the abscissas are logarithms of the total numbers of atoms.

corresponds to a given value of b. Figure 15 shows a number of theoretical curves of growth, and Figure 16 gives a series of observed curves plotted within the framework of the theoretical curves. Finally in Table III we have given the relevant quantities for seven different stars. The column labeled "fictitious temperature" shows what T would have to be in order to bring about agreement of the observational curve with the theoretical curve. Clearly, these temperatures are not compatible with the effective temperatures or with the excitation temperatures determined from the spectra of the first three stars. For the last three stars, which are members of the main sequence, thermal motions will account for the observed curves of growth. In

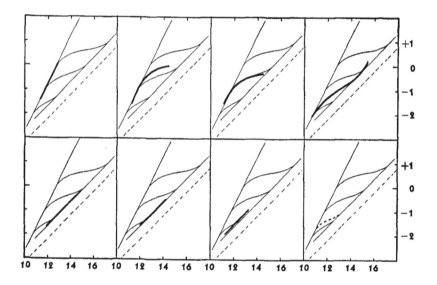

FIGURE 16. Observed curves of growth for seven stars (heavy lines). Upper row: 17 Leporis, Epsilon Aurigae, Alpha Persei, the Sun. Lower row: Alpha Cygni, Alpha Canis Majoris, Alpha Lyrae. The thin continuous lines represent a framework of theoretical curves with three different values of b, as in Figure 15 and with a damping constant which is ten times the classical value. The coordinates are the same as in Figure 15.

the case of Alpha Cygni there is a strange lack of turbulence inferred from the curve of growth. Yet this star is a supergiant and the shape of its lines strongly suggests that we are dealing with a highly turbulent reversing layer. We have as yet not succeeded in explaining this discrepancy. But in several other stars there appear to be indications that the turbulent elements are relatively large compared to the thickness of the absorbing layer; hence in such stars the curve of growth does not give a reliable determination of the velocity distribution of the turbulent elements. The spectral lines are broader than would be consistent with the usual turbulent velocities. For example, in Delta Canis Majoris some of the strongest absorption lines have widths which require velocities of the order of about 50 km/sec. Yet the curve of growth determined by Miss H. Steel suggests a turbulent velocity of only about 5 km/sec. A similar result was obtained by M. Schwarzschild for Eta Aquilae and by K. O. Wright and Elsa Van Dien for Epsilon Aurigae. In both stars the contours of the absorption lines are too broad. As Wright has suggested, such a broadening could be produced by axial rotation, but perhaps it is more reasonable to suppose that the turbulent elements are so large that there is no complete statistical superposition of the random motions of the individual elements, and a kind of convection broadening is combined with the regular turbulence broadening.

TABLE III

Turbulence in Stellar Atmospheres

Star	b	v_0 (Turbulent Velocity or Thermal Velocity)	T (Fictitious Temperature or Effective Temperature)
17 Leporis	1.0 Å	67 km/sec	3×10^7
Epsilon Aurigae	0.3	20	2×10^6
Alpha Persei	0.1	7	3×10^5
Alpha Cygni	0.03	2	10^4
Alpha Canis Majoris	0.03	2	10^4
Alpha Lyrae	0.03	2	10^4
Sun	0.03	2	10^4

Finally, we must remember that the assumption of a Maxwellian distribution is only justified by the agreement which we have reached in fitting the observations to the theoretical curves of growth. There is no good physical reason for supposing that the distribution of velocities must of necessity be Maxwellian. It is much more probable that we are concerned with the absorption within a vast field of stellar prominences whose motions are probably primarily directed radially from and toward the star's center. If we think for a moment of the prominences of the sun we should expect a tendency to show at the center motions of approach and of recession, predominantly. This may be the case in some other stars. For example, in supergiants of late spectral type, W. S. Adams has found double absorption components in some of the low-level lines. Such double lines have been found, for example, in Alpha Scorpii and Alpha Orionis. Interpreted in terms of turbulent motions they would indicate the presence of a large number of prominences which, predominantly, move away from the star and another large group of prominences which move in the direction of the star's center. There are also prominences whose velocities are tangential with respect to the observer, and they can be produced not only at the star's limb but also near the center of the disk whenever a prominence happens to be moving at right angles to the star's radius. The number of possible distributions of velocities is of course infinite, and there is no possibility of solving for the distribution when we have at our disposal only the observed curve of growth. An infinite variety of solutions would be possible, but it is well to remember that the Maxwellian distribution is only a guess, so that it may be futile to engage in highly refined computations as long as this crucial question remains unanswered.

The theory of turbulence predicts a law of velocities depending upon the size of the turbulent elements, and there are indications that such a distribution accords with the observations. It has also been suggested that in super-giants (Delta Canis Majoris) the turbulent velocities increase with the height in the atmosphere. These ideas may be consistent with one another. But until this question has been resolved the observed curves of growth can give only rough approximations to the abundances of the various atoms.

The most recent work on the abundances in normal stars of the main sequence and of the giant branch is that of K. O. Wright. His values are in general very similar to those compiled by Unsöld except for a constant quantity which depends upon the optical thickness of the reversing layer down to the imaginary layer which we designate as the photosphere. In Wright's work the level of the photosphere was assumed to be at an optical depth of $\tau = 0.3$, so that in all the stars considered the sums represent the actual amount of each element in a column of cross section one square centimeter above that level. Wright points out that there are about ten times as many atoms above this particular level in Gamma Cygni as there are in the sun. This would confirm what had been believed previously, namely that the atmospheres of the giants are much more extensive than those of the dwarfs. There are no striking differences between any of the abundances listed; apparently the relative composition of Alpha Canis Minoris (Procyon) is identical with that of the sun and the composition of Gamma Cygni is nearly identical with that of Alpha Persei. However, it seems that sodium and yttrium may be more abundant in the giant stars than in the dwarfs, while magnesium and silicon may be less abundant. The uncertainty of this work is estimated as being of the order of approximately 0.5 in log N. Wright's results are summarized in Table IV.

A considerable degree of uncertainty remains with respect to the theory of the formation of an absorption line. Another serious difficulty arises when we are concerned with lines of large excitation potential for which an incorrect assumption of the temperature may easily produce an error in the resulting atomic abundance by a factor of several hundred. Hence several recent investigations have been devoted to the study of lines in which the Boltzmann factor is small. With regard to the problem of radiative transfer, we have used on page 70 the case of a simple exponential absorption without re-emission. This would be correct only for absorption in a gaseous medium which is far removed from the source, such as a shell or ring,

TABLE IV

Number of Atoms in Solar-type Stars above Level $\tau = 0.3$

Element	Sun log N	Gamma Cygni log N	Alpha Persei log N	Alpha Canis Minoris log N
C	19.46	20.22	19.62	19.48
Na	17.62	19.13	18.62	17.72
Mg	20.31	20.77	20.41	20.36
Al	17.22	18.55	17.88	17.20
Si	18.79	19.36	19.27	18.84
Ca	17.54	18.72	18.53	17.72
Sc	14.18	15.36	15.06	14.25
Ti	16.15	17.26	16.89	16.15
V	15.14	16.37	16.46	15.20
Cr	16.77	17.72	17.37	16.86
Mn	17.18	17.90	17.68	17.21
Fe	19.03	20.08	19.76	19.18
Co	17.17	17.92	17.91	17.45
Ni	18.41	19.37	19.17	18.48
Cu	17.62	18.39	17.92	17.77
Zn	18.16	18.66	18.38	18.39
Sr	13.81	14.80	14.88	13.92
Y	13.10	14.64	14.18	13.40
Zr	13.97	15.07	14.43	13.89
Ba	13.34	14.28	14.06	13.33
La	12.32	13.33	12.87	12.17

or a nebula. In a stellar reversing layer the re-emitted radiation is added to the outgoing flux of radiation; hence the intensity which we observe is greater than in the exponential case. M. Minnaert, and later A. Unsöld, remedied this by means of a semi-empirical formula for the contour of an absorption line, which agrees reasonably well with the theory of radiative transfer. The work of S. Chandrasekhar has given an exact theory of radiative transfer in the Milne-

Eddington model. It is no longer necessary to introduce artificially a fictitious photosphere at some level such as $\tau = 0.3$. Instead, we define η as the ratio of the absorption coefficient in the line to the continuous absorption coefficient, and we assume that this ratio is constant throughout the atmosphere. We can then express the Planck formula for the radiation in terms of a linear approximation as a function of the optical depth, and compute accurate values of the equivalent widths. Such computations have been carried out by M. H. Wrubel for different values of the linear expansion of the Planck function and for different values of the Doppler broadening. The resulting curves differ appreciably from those which have been computed by Unsöld with the help of the semi-empirical solution of the radiative transfer suggested by Minnaert.

The seriousness of this problem has been emphasized by L. H. Aller who has investigated the curves of growth of the B-type star Gamma Pegasi. He obtained appreciable differences depending upon whether he used Unsöld's method or Strömgren's theoretical curve of growth computed numerically. For the ratio of hydrogen to oxygen in Gamma Pegasi, a star of spectral type B2, he finds 10,000, which is ten times larger than Unsöld's value for Tau Scorpii. Yet Aller is not certain that this difference is real.

At the present time we can state that the study of the abundances in normal stars of the main sequence and in giants has led to the establishment of what might be called the original composition of galactic matter. Perhaps the most striking result of all the discussions is the remarkable degree of uniformity that is observed in the most widely different astronomical sources. The sun, the main-sequence stars of type F, the helium stars like Tau Scorpii, and even the O-type star 10 Lacertae, all have approximately the same composition. The observers estimate that the precision with which the abundances can now be determined corresponds to a factor of two or three. Even more surprising is the fact that Strömgren's result for interstellar matter and D. H. Menzel's for the planetary nebulae also indicate a composition that is essentially the same as that of the normal stars. This is also confirmed by the work of I. S. Bowen and A. B. Wyse on diffuse emitting nebulae. As we have already seen, the work of Schwarzschild and especially of Mrs. Harrison has led to essentially the same composition of the stellar interior, namely, about 59 per cent by weight of hydrogen, 30 per cent by weight of helium, 9 per cent of the oxygen group, and 2 per cent of the metals.

A critical study by Harrison Brown of the abundances in stars and meteorites has led to the determination of the uncertainties of the observational results for a long list of elements. These values indicate that, with the exception of the volatile gases, the composition is probably the same. Only strontium and zinc are slightly outside the range permitted by the uncertainties, in the sense that both are a little more abundant in the stars, but it is probable that the stellar uncertainties for these elements have been underestimated.

11. Anomalous Abundances

Our next problem is to consider possible departures from these normal abundances. We are here encountering some disadvantages, but also some decided gains. The stars which we suspect of having unusual abundances are often faint and therefore not within the reach of our most powerful spectrographs. Moreover, few of them are binaries, so we cannot determine their masses, and only a few are sufficiently well known to be placed accurately with respect to their luminosity. But there are two important advantages. The first is that we do not have to determine absolute abundances, but may be satisfied with differential measurements against normal stars or against other stars of known characteristics. The second consists in the fact that the spectroscopic peculiarities which we shall attempt to describe in terms of chemical composition are often exceedingly large so that even without accurate measurements we can draw important conclusions concerning the absence or presence of certain atoms.

Perhaps the best procedure will be to examine the H-R diagram and to discuss separate groups of stars with relation to their place in that diagram.

We have already commented upon the subdwarfs, for which Morgan found that those of earlier type tend to have weak hydrogen and weak metallic lines, while those of later type are principally characterized by the strength of certain molecular bands, especially that of CH at λ 4300. These stars are representatives of Baade's Population II, which is characteristic of the central regions of our galaxy and also of the globular clusters. The horizontal branch of Population II contains a number of interesting objects, especially the cluster-type variables, among which RR Lyrae is the best-known example. These high-velocity cluster-type variables have anomalous spectral types. Not only do they vary in type, as do the normal classical Cepheid variables of Population I, but they are peculiar because of the faintness of their hydrogen absorption lines compared with the lines of ionized calcium. The first impression upon looking at the spectrum of a cluster-type variable is that the hydrogen lines require a later spectral type than the lines of ionized calcium. The rest of the metallic lines seem to agree with those of ionized calcium. Thus in RR Lyrae at minimum light the hydrogen lines indicate a spectral type of F6 and the metallic lines indicate F0. At maximum light the hydrogen lines indicate a type of F0 and the metallic lines a type of A2. Numerous observations of other cluster-type variables have shown that this phenomenon is a general one, and it is not possible to fit them within the framework of the usual spectral classification. The latter was essentially developed from the stars of Population I. G. Münch and L. R. Terrazas have found that the colors of the cluster-type variables agree with the spectral types determined from the metallic lines, so we must consider the intensities of the hydrogen lines as being anomalous. The temptation is great to explain the anomaly of the H lines in the stars of Population II in terms of hydrogen abundances. The tendency in many high-velocity stars seems to be the weakness of the hydrogen lines, and this induced Kuiper to suggest that the stars of Population II may have a relatively small abundance of hydrogen, so that "on the whole the hydrogen content increases with decreasing ellipticity of the galactic orbits." This would be in agreement with observations of high-velocity A-type stars in which the hydrogen lines are sharper and the metallic lines weaker than in normal stars of the same spectral class. But it is difficult to reconcile the spectroscopic features of Population II with the requirements of the theory. If hydrogen were relatively less abundant, this would be reflected principally in the continuous absorption. Among the early-type stars, where atomic hydrogen is responsible for the continuous absorption, a reduction will simply mean that in the place of the Paschen, Balmer, and Lyman continuous absorptions we shall primarily be concerned with absorptions produced by helium and the other light elements, such as oxygen and nitrogen. Assuming that the absorbing power of these elements is about the same as that of hydrogen, the principal result should be a weakening of the Balmer absorption lines; however, as Greenstein has shown, the substitution of helium for hydrogen should make the atmosphere more transparent and strengthen the lines of the metals and of helium. What would appear to be rather difficult to explain is the tendency in many stars of the high-velocity group to have weak lines of all elements, including the metals. Why that should be is not easy to explain. It would imply that the place of hydrogen in the reversing layers of these stars is occupied by some agent which is much more efficient in producing continuous absorption than is atomic hydrogen or the negative hydrogen atom. Another possibility is that these stars have a thin atmosphere because of high surface gravity.

In the high-velocity giants of types G and K, P. C. Keenan, W. W. Morgan, and G. Münch found a marked weakening of CN, while the H lines and the CH band are normal or slightly stronger than in ordinary stars. In the R-type high-velocity stars, which have strong molecular bands involving C, the CH band at λ 4300 shows "such

tremendous strengthening that they have been called CH stars."
Here the H lines are slightly strengthened. The molecular features
are more affected than the atomic lines and the light abundant elements H, C, N and O are the ones mostly involved. Thus, the criteria are somewhat conflicting, compared with those we discussed
before.

We have ignored the change in luminosity which may be the result
of the depletion of hydrogen in the stellar interior. That is altogether
a different matter, and its observed characteristics depend upon
whether or not the stellar material is completely mixed.

But there is another difficulty, that of the ages of the stars. Low
hydrogen abundance in subdwarfs of the solar type would imply a
difference in the original composition of the medium out of which
the stars had condensed. The reason for this is that nuclear processes
in a G-type subdwarf could not have produced an appreciable change
of composition during the lifetime of the galaxy of 3×10^9 years.

But the greatest difficulty in explaining the characteristics of Population II in terms of lowered hydrogen abundance arises from the fact
that the dwarfs fall below the main sequence by about one or two
magnitudes. This is difficult to understand if these stars make use of
the Bethe cycle for their energy production. Hence we must at present
admit that our information is not sufficient to establish beyond doubt
the composition of the stars of Population II. The fact that the O, B,
and early A stars are rare in that population may be the result of their
conversion by nuclear processes of hydrogen into helium. But the
tendency of the subdwarfs to be fainter than the dwarfs of Population
I has at present no logical explanation.

There are within the regular Population I a number of small groups
of stars which fit into the principal branches of the H-R diagram but
show certain anomalies with respect to their chemical composition.
Starting with the left side of the H-R diagram, we have first the very
striking split of the Wolf-Rayet stars into two groups: the carbon
sequence and the nitrogen sequence. These stars are extremely rare.
Only about one hundred are known at the present time and the total
number within our galaxy probably does not exceed a few thousand.
They are apparently members of Population I and do not occur in
the central bulge of the galaxy. Their spectra are characterized by
strong and extremely broad emission features of highly ionized atoms
which are superposed over continuous spectra whose energy distribution is that of a very hot star, perhaps of temperature 20,000° or
30,000°. The emission bands have been identified mostly through
the work of B. Edlén in Sweden and are now known to be caused
principally by the atoms of He I, He II, N III, N IV, N V, O III, O IV,
O V, O VI, Si II, Si III, Si IV, C II, C III, and C IV. There are no forbid-

den transitions. Both branches of the Wolf-Rayet stars contain He I, He II, and H. One sequence has strong lines of C II, C III, and C IV, while the other branch has strong lines of N III, N IV, and N V. However, P. Swings has shown that in some stars of the nitrogen sequence the C IV doublet λ 5802–5812 is as strong as is the D_3 emission of He I. He has also shown that certain nuclei of planetary nebulae are as rich in N as in C. Both sequences can be tied in with the main sequence of the H-R diagram in the region of the O-type stars, or somewhat below it. The criterion which is used for the classification of the Wolf-Rayet stars is the ratio He II 5411/He I 5875. The Wolf-Rayet stars show few, if any, absorption features. There are violet absorption borders in connection with some of the broad emission bands and a few of the stars have weak and somewhat diffuse central absorption features in the emission bands. C. S. Beals has found an interesting relation between the average widths of the bands and the spectral type. As a rule the earlier spectral types, those which are characterized by the higher amount of ionization, are at the same time those which have the broadest bands. There is also a well-defined relation for each star between the widths of the bands and the corresponding ionization potentials. For example, in the star HD 192163, of spectral type WN6, the widths of the emission bands of He I corresponding to an ionization potential of 24.5 volts average 1815 km/sec, while those of N v corresponding to 97.4 volts average 487 km/sec.

This remarkable relationship must be due to stratification within a tenuous mass of gas surrounding these hot stars. Following Beals, we explain the great widths of the emission bands as being caused by the expansion of the shell of gas, so that the increase in expansion velocity with decrease of ionization potential probably implies outward acceleration of the gases, because we believe that the exciting high-energy ultraviolet radiation is effective only in the inner portions of the shell and becomes exhausted relatively close to the star. Hence the emission features of N v presumably come from an inner layer in the shell, while the emission bands of He I come from an outer layer. This picture is similar to that proposed by I. S. Bowen for the planetary nebulae. Some doubts have been expressed in recent years concerning the expansion theory of the emission bands in Wolf-Rayet stars. It has, for example, been suggested that the very large widths of these features may be the result of electron scattering in a vast shell of free electrons kept neutral by the presence of a relatively small number of completely or nearly completely stripped nuclei. Because of the small mass of an electron, its average thermal motion is very much larger than that of an ion. Hence, if scattering by free electrons is predominant, a relatively narrow emission line produced by the atoms of, let us say, N v might well be broadened until it is similar to the features observed in the Wolf-Rayet stars.

In principle it is a relatively easy task to measure the intensities of the emission features and to determine from them the numbers of atoms present in the upper states of the transitions represented in our spectra. It is true that we do not possess accurate transition probabilities for the atoms under consideration, but approximations based upon the transition probabilities in the hydrogen atom cannot be too far wrong if adapted to these high-level lines of highly ionized systems. Hence it is impossible to avoid the conclusion that carbon is very abundant in the one group of Wolf-Rayet stars while nitrogen is very abundant in the other. At the same time H, He, Si, etc., seem to be about the same in both groups.

There is at present no plausible explanation of the differences just noted, but we must remember that in the case of the Wolf-Rayet stars we are concerned with an even more ephemeral structure than in the case of the reversing layers of main-sequence stars. We do not even know how these highly ionized shells have been produced and why they are present in a very small number of O-type stars and are not present in all other stars occupying the same location in the H-R diagram.

The Wolf-Rayet stars are peculiar in several other respects. Most of them are members of binary systems in which the other component is an ordinary absorption O or B star. At one time it seemed probable that all Wolf-Rayet stars were members of binary systems, but several are now known which show no appreciable change in radial velocity and therefore are not ordinary spectroscopic binaries. They may, however, be double stars of longer periods, or they may be close systems whose orbital planes are at right angles to the line of sight.

Considered as binary systems, these objects are also somewhat unusual. The Wolf-Rayet components are invariably the more luminous members of the systems, but their masses are roughly only one-half as large as the masses of their normal absorption-type companions. Hence, the Wolf-Rayet stars appear to violate the mass-luminosity relation.

It would be unwise at this stage to attribute any great physical importance to the differences of composition in the Wolf-Rayet group. Probably the ionized shells have been produced by a process which is the result of binary structure. It is possible that selective effects are present in such systems and that they have nothing whatever in common with the nuclear processes going on within the stellar interiors. It has sometimes been suggested that the Wolf-Rayet stars are deficient in H, but P. Swings has recently concluded that the intensities of the Balmer lines in emission may be quite normal.

If we descend along the main sequence we encounter next a few strange O-type stars with varying kinds of emission features, principally those of He II 4686, N III 4641–4643, and C III 5686. These stars are designated as Of stars and they seem to be quite frequent among the ordinary absorption O's. In fact, the more detailed investigations of chemical composition which have been made to date refer to stars, such as 10 Lacertae, which have only absorption lines and closely resemble the normal B-type stars except for a higher degree of ionization. Although 10 Lacertae would usually be described as a normal O-type star, this kind of object is relatively infrequent in the galaxy. Ordinarily, an O-type star has broad spectral absorption lines with contours suggesting that the broadening is not entirely due to rotation but is often produced by a large amount of turbulence. Many of these O-type stars have emission features at He II 4686 and N III 4641–4643. Occasionally the H lines are bright. There seems to be a fairly uniform transition from the relatively rare peculiar absorption O stars to the peculiar Of stars in which the emission features are conspicuous. The latter stars have been investigated primarily by Swings and Struve. Usually their spectra are variable. The intensities of the emission lines vary in an irregular manner, and occasionally violet absorption borders are seen in connection with the bright lines. The brighter component of the spectroscopic binary 29 Canis Majoris is such a variable Of star.

It is probable that the bright lines in the Of stars are related to the broad emission features in the Wolf-Rayet stars, but in the former they are much more narrow, their widths corresponding to some 50 to 100 km/sec only. Among the Of stars there seems to be no clear-cut splitting into a carbon and a nitrogen branch. On the other hand, Swings and the author have shown that there are some peculiar selectivities among the emission lines: for example He II 4686 is often an emission line, although all other He II lines are present only in absorption. There are other similar selectivities among the lines of C, N, etc. The processes of excitation are not simply those which can be described by the Boltzmann relation. The conditions must be far removed from those of thermodynamical equilibrium, and the populations in the different energy levels of the atoms are built up through the operation of various radiative processes of excitation. It is therefore clear that lines which are produced in this manner cannot be relied upon to give estimates of chemical composition, unless we know exactly the precise mechanism by which the populations are built up. Otherwise, we should obtain contradictory results from different lines.

Although the normal pure-absorption O-type spectra are relatively rare, there is no reason at present to suspect that the chemical com-

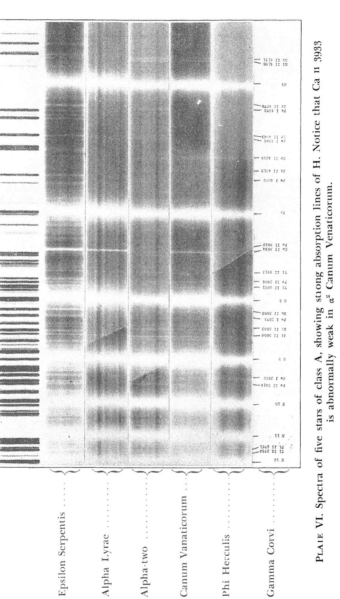

PLATE VI. Spectra of five stars of class A, showing strong absorption lines of H. Notice that Ca II 3933 is abnormally weak in a^2 Canum Venaticorum.

Epsilon Serpentis

Alpha Lyrae

Alpha-two

Canum Vanaticorum

Phi Herculis

Gamma Corvi

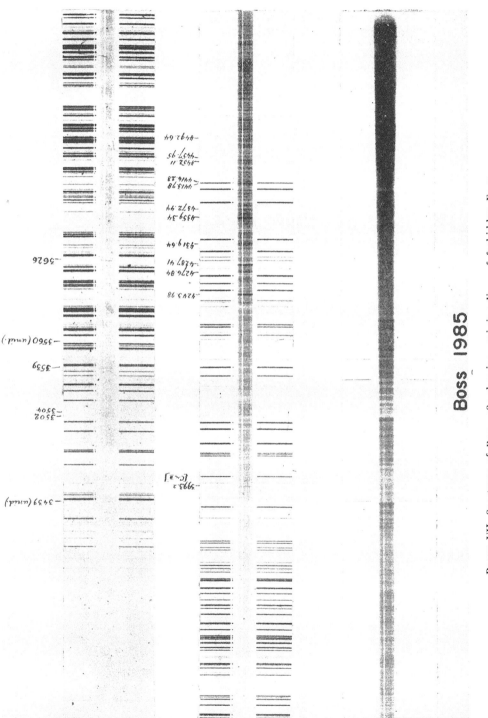

Boss 1985

PLATE VII Spectrum of Boss 1985 showing emission lines of forbidden Fe II.

position of the emission-type Of stars is different from the normal composition discussed previously.

Next in line along the main sequence are the B-type stars. They are characterized primarily by a surprising degree of uniformity in spectral characteristics. The principal variation among the stars of this class arises through differences in stellar rotation. We now know that the B-type, and some of the O-type, stars often possess very large velocities of rotation. We also know that these velocities are not the same in all stars. Some stars, like 10 Lacertae or Tau Scorpii, have no measurable rotation, while others have rotational velocities of the order of 400 or 500 km/sec. The largest rotational velocities occur among certain single stars and also among some long-period spectroscopic binaries, like Phi Persei and Zeta Tauri—which have at the same time bright lines of hydrogen and sometimes of Fe II. The effect of rotation causes the spectral lines to be diffuse and faint when examined with the eye, on a spectrogram. Hence it is sometimes difficult to distinguish between very large rotational broadening and a true physical weakening of the lines. Nevertheless, it is now known that B-type stars exist in which all lines are weaker than normal. The brighter B-type stars in the Pleiades belong to this group, and there are other examples throughout our galaxy. This general tendency towards weak absorption lines is similar to the weakening of the lines in some stars of Population II, and the cause may also be the same.

It is of interest that the rotational velocities of the components of close spectroscopic binaries never show excessively large values. Equatorial velocities of the order of 250 km/sec, as in Alpha Virginis, are never exceeded (at least among the systems for which we have adequate observational data). It is probable that in the formation of a binary out of a single star (or in the reverse process) the total angular momentum is conserved. Since, in a binary, part of the momentum goes into orbital motion, the momentum available for the axial rotation of each component is greatly reduced.

A few B stars of the emission type have been shown by Morgan to contain spectral lines of hydrogen whose absorption wings are much stronger and broader than is normal in ordinary absorption stars of similar spectral subdivisions. These stars have narrow emission lines of hydrogen and also narrow absorption lines of helium and other elements. Morgan believes that they are rapidly rotating B-type stars surrounded by emitting rings of gas oriented in such a way that their axes of rotation lie close to the line of sight. The great width and strength of the hydrogen absorption lines indicate excessively large Stark effect in the polar regions of the rapidly rotating and, therefore, highly flattened B stars. Hence this phenomenon is probably also without significance from the point of view of chemical composition.

Combining all the evidence we have from the B-type stars we should be inclined to suspect only those objects which have generally weak absorption lines of all elements. Otherwise, the chemical composition is remarkably uniform.

Before we go on, we must qualify this statement. There is one B star, HD 124448, which was found by D. M. Popper to have almost no absorption lines and no emission lines of hydrogen, but, instead, a particularly strong set of all the lines of He I. Not only are the Balmer lines absent, but the hydrogen discontinuity at λ 3647 is also absent. The helium lines, on the contrary, are stronger than those observed in any other known star. All other elements which might be expected in a B2 spectrum are present. Popper has observed faint discontinuities corresponding to the ionization limits of the various He I series. These discontinuities are about as strong as would be expected if the reversing layer consisted entirely of helium. His observations lend strong support to the hypothesis of an atmosphere without hydrogen, because any appreciable admixture of the lighter gas would greatly reduce the calculated discontinuity of He I. Of course the question immediately arises whether this phenomenon is unique, and whether it is confined to the outer layers of the star or is a consequence either of a deficiency of hydrogen when this star was formed, or of the action of nuclear processes in the star's interior.

The very great uniformity in the composition of the vast majority of the B-type stars, and the similarity of this composition to that of the sun, suggests that either there is no appreciable mixing or we happen to be observing only young stars, and rarely if ever catch them in the transition stage at which hydrogen has been sufficiently exhausted to change the spectrum, but not enough to move the star entirely beyond the region of the B stars in the H-R diagram. As yet we have very little information concerning the other physical characteristics of HD 124448. Popper estimates that the visual absolute magnitude is between -2.5 and -4.0. This agrees with the general character of the spectrum in showing that the star is not a supergiant. It is probably closely related to the star Upsilon Sagittarii (Plate V) which has been investigated by several astronomers, most recently by J. L. Greenstein. It is usually classified as an A star; but it, too, has only extremely weak hydrogen lines, while all the other lines, namely those of the singly-ionized metals as well as those of He I, N I, N II, Mg II, Si II, and S II, are very strong. There are present even lines of such high stages of ionization and excitation as those of Si III and Fe III. Greenstein has reached the conclusion that if the equivalent widths of the spectral lines are interpreted in the usual manner with the help of the curve of growth, the abundance of hydrogen is found to be quite small. Helium is about 100 times as abundant as iron. The presence of lines of high excitation potential, notably of A II, indicates a high excitation temperature for the extreme ultraviolet region. But the small abundance of hydrogen does not depend upon a departure from thermodynamic equilibrium in the distribution of energy of the exciting radiation. The point here is that hydrogen is weak, not only with respect to lines of higher excitation and ionization, but also with respect to lines of much lower excitation and ionization. Again, there is no sound reason for doubting the direct evidence of the spectra: hydrogen must be deficient in the reversing layer of Upsilon Sagittarii. At the same time, this spectrum gives strong indication of very high luminosity. The absolute magnitude suggested by Greenstein is −7.

Still another star of the same kind was found by W. P. Bidelman; its name is HD 30353. This star is somewhat later in type and is therefore cooler than Upsilon Sagittarii, but it shares with the latter the striking weakness of the hydrogen absorption lines and the great strength of the metallic lines as well as those of He I. It is possible that this star forms a transition between Upsilon Sagittarii and Popper's star, on one side, and the stars of the R Coronae Borealis type, which are also characterized by extremely weak lines of hydrogen, on the other. The spectrum of R Coronae Borealis near maximum light was investigated by L. Berman, who constructed a dependable curve of growth and determined the abundances of the elements in the usual manner. He found that carbon accounted for 69 per cent by volume, H for 27 per cent, Mg for 2 per cent, N for 0.3 per cent, and Fe I for 1 per cent. The rest of the elements account for much smaller fractions of the volume. It is probably not reasonable to regard R Coronae Borealis as physically similar to the stars of constant light investigated by Popper, Greenstein, and Bidelman. R Coronae Borealis is an irregular variable whose light changes between apparent magnitude 6 and apparent magnitude 14. When the spectrum is very faint its character is entirely different from that observed by Berman. The absorption lines are very weak and hazy. Instead, emission lines of Ca II, Sc II and probably He I are conspicuous, but there are no appreciable emission lines of hydrogen. Hence, it is reasonable to suppose that the emitted gases are also deficient in hydrogen when the star is at minimum light.

To these stars may be added several other objects of the R Coronae Borealis type, for example HD 25878, which was investigated by Bidelman. All of them are so peculiar that it is difficult to classify them accurately with respect to absolute magnitude. In all probability they are rather luminous stars, although Popper's B-type star is not a supergiant, as we have seen.

As in several previous instances, the evidence seems to be fairly reliable that hydrogen is much less abundant in the outer layers of these stars than in stars having normal spectra, but again we fail to recognize any consistent connecting features which would permit us to attribute these anomalies of composition to a single physical process, such as nuclear transformation. It is significant that Upsilon Sagittarii, as well as HD 30353, are spectroscopic binaries of long period. The former has been known for a long time. The latter was recently found by Bidelman to have a range in velocity of something like 100 km/sec, with a period of the order of one year. The combination of this range in velocity and length of period imply that the mass function must be very large. The binary must be massive, but otherwise the character of the orbit remains unknown. It is tempting to believe that in some unknown manner binary nature rather than nuclear processes are responsible for these anomalies of composition.

Returning to the main sequence, we next have the stars of spectral class A. They have always formed a particularly difficult group of objects for persons engaged in the classification of stellar spectra; and even though the H-R diagram gives no indication of it, the sequence is really divided into several approximately parallel and fairly numerous groups which differ only slightly in absolute magnitude from one another. Many years ago it was found that among the A-type stars there are some in which the lines of Si II are particularly strong, and it was possible to classify these Si II stars in a sequence running parallel to the ordinary main sequence. Similarly, there are some stars in which the lines of Sr II or Cr II are unusually strong, and also stars in which the lines of Mn II are abnormally intense. A typical example of this latter group is Alpha Andromedae; a member of the Si II group is the star Tau-nine Eridani. An example of the Cr II star is 73 Draconis. W. W. Morgan has shown that among the A-type stars the usual two-dimensional classification recorded in the H-R diagram is not sufficient. He concludes that even three dimensions may be insufficient for a general classification. Moreover, the dispersion in the intensities recorded for lines of different elements is not confined to those elements whose behavior had been recorded as peculiar on spectrograms of low dispersion. Any complete scheme of classification of the A-type stars would be almost impossibly complex; but in spite of the absence of order in the behavior of the elements, there are definite indications that at least one other parameter is needed in addition to the temperature and surface gravity (or pressure). Many years ago Morgan regarded it as probable that this parameter is the abundance of each element. But it must be recognized that there is no good physical cause that would explain these apparent differences in the abundances of certain elements. For example, it is not easily understandable why in some stars Si should be relatively more abundant by a factor of perhaps 10, or even 100, than it is in the normal sequence defined by such stars as Alpha Lyrae and Sirius. It is especially puzzling that we should have this confusion only among the A-type and early F-type stars, but not among the cooler stars nor among those of classes B and O. One possibility would be to suppose that the A-type stars are old enough to show changes in composition and are not developing rapidly enough to remove them quickly out of the domain in which they are located in the H-R diagram, while the O-type and B-type stars develop so rapidly that we observe them only during a very short interval of their existence before nuclear processes have greatly altered their composition; but this explanation is not very satisfactory. Moreover, there is nothing in the Bethe process that would give us anything like the observed variety in the compositions of the heavier elements.

It is perhaps significant that these peculiar sequences are all concentrated in the region of temperatures where the Balmer lines of H are at their maximum of intensity. The elements which behave in a peculiar manner all have second ionization potentials which are not far removed from 13.5 volts, the ionization potential of H. Hence, it is possible that differences in the continuous spectra of the stars in the region of λ 911—the Lyman limit—should have a bearing upon the ionization of the anomalous elements.

A remarkable group of class A are the so-called metallic-line stars. These stars were discovered by Morgan and his associates at the Yerkes Observatory and include a number of objects which had previously been assigned to one of the other groups of peculiar A stars. As a rule the metallic-line stars give contradictory spectral classes when different criteria are used. The metallic lines are strong enough to place them among the F stars. The H lines, however, are also strong, and require that they should be placed among the A stars. The most recent investigation of this large group was made by J. L. Greenstein. If the intensities of the spectral lines are discussed and interpreted in the conventional manner, there is no escape from the conclusion that certain elements, such as Ca, H, Sc, Zr, and Mg, have abundance deficiencies by a factor of 10—a large quantity when it is translated into observed intensities of the spectral lines. Greenstein has struggled with this problem for a long time. Why should it be that these particular elements are deficient? There is apparently nothing in the atomic numbers or in the other physical or nuclear properties of these elements that would suggest an explanation. After a long search, Greenstein finally discovered that when the observed deficiencies were plotted against the second ionization potentials of the elements, a uniform curve was obtained. All those elements whose second ionization potentials are between 11.8 and 16 volts are deficient, and no others. This range of potentials is close to the ionization potential of hydrogen, 13.5 volts. Greenstein correctly suggested that in all probability there is a relation between the observed deficiencies and the ionization of hydrogen. That much is reasonably certain. But the exact physical mechanism is not yet certain. One possibility would be that the second stage of ionization is greatly enhanced, so that there are fewer atoms available to produce the absorption lines of Ca II, Sc II, etc. If, for example, there were in the vicinity of a metallic-line star a gaseous medium which is rich in Lyman emission lines, the radiation required for the production of strong second-stage ionization in the elements under consideration would be increased. This, in turn, would mean that there would be weaker lines of precisely those atoms in which deficiencies were observed. But Greenstein was not entirely satisfied with this mechanism, because similar considerations had failed to lead to satisfactory results in several other types of stars. He called attention to the possibility that certain effects of collisions could, through a process of resonance, result in the removal by a proton of an electron from a singly ionized metal, thereby causing an excessive amount of second-stage ionization in those metals whose second ionization potentials are close to 13.5 volts. Greenstein also found that the metallic-line stars show unexpectedly large turbulent velocities and low surface gravity. As a result, these stars are overluminous and may perhaps be deficient in hydrogen, as we had already suggested in our discussion of the A stars in the Hyades.

Distinct from the metallic-line stars is a group of A stars which are characterized by strong lines of the rare earths. Some of them may at the same time show other anomalous line-intensities, for example, strong Si II and weak Ca II. The first star of this kind which came to the attention of astronomers was Alpha Canum Venaticorum, or Cor Caroli (Plate VI). Many years ago A. A. Belopolsky discovered that in this star the spectral lines undergo periodic changes in intensity and that one group of lines becomes very strong when the other group becomes faint. The period of variation was found to be 5.50 days, and this periodicity is as regular as in a spectroscopic binary. But despite some early announcements that the star is a spectroscopic binary, more recent investigations have shown that the changes in the wave lengths of the lines are complex and cannot be interpreted in terms of binary motion. The most astonishing feature of this star is the intensity of the rare earths, especially Eu II. Within the 5.5 day period, all the lines of the rare earths change in intensity, becoming very strong at times and disappearing or almost disappearing at others. The range of intensity variation is not the same for all the rare earths, Eu II showing the greatest change, with Dy II and Gd II following next. The second group of lines, which changes in the opposite direction, is illustrated principally by the element Cr II, but the range of the intensity-variation is smaller than for the rare earths. Finally, there are a number of elements which change little or not at all. Among them are Si II and Mg II. As the intensities of the spectral lines undergo their periodic changes, the radial velocities derived from them also vary with the same period. The lines of the rare earths, and a few other elements predominantly of low ionization potential, show a shallow minimum of radial velocity when these lines are weak, and a sharp maximum when they are strong. The lines of Cr II and of several other elements have a double wave in their velocity-curves, maximum occurring approximately at that phase at which the rare earths have minimum radial velocity. Finally, those elements which show little or no variation in intensity have constant radial velocities. Among them are Mg II, Si II, H, and Ca II. The variations in radial velocity cannot be attributed to binary motion and must be caused by internal motions in the absorbing layers of the star. The contours of the lines are not uniformly sharp but undergo periodic variations in width. The lines of Cr II seem to change in such a way that they are broadest when the corresponding radial velocity is largest. There are two maxima in the curve of line-widths, corresponding to the two maxima of the radial velocity.

There is a general tendency for the elements of the two groups to have different ionization potentials. This fact was used by W. S. Tai in an attempt to explain the variations of the line intensities in terms of the ordinary ionization theory. However, this attempt was not successful; in fact it tended to obscure the very important fact that the classical theory of line intensities will not account for the observed amount in the change of intensity nor for the great strength of the europium lines when they are near maximum intensity. W. A. Hiltner found that at certain times the lines of Fe II and Cr II became not only broad but double. These duplicities are, however, in no way related to those duplicities which are sometimes observed in a spectroscopic binary. Their nature has not been explained. In all probability they must have something to do with turbulent motions in the atmospheres of these strange stars.

Through the work of W. W. Morgan and A. J. Deutsch, numerous other spectrum variables of type A have been discovered. Many but not all contain the rare earths with unusual intensities. In some the variations are mostly confined to ordinary elements found in all A-type stars, such as Cr II. In addition, Morgan, and later Hiltner, have investigated several stars in which the intensities of the lines are constant but in which the rare earths are also unusually strong. Among these, Beta Coronae Borealis is outstanding.

It would have been reasonable on the basis of straightforward curve-of-growth investigations to attribute exceptionally large abundances to the rare earths in these stars, exceeding those observed in the normal A stars by factors ranging from 10 to perhaps 1,000, and we would even be compelled to believe that in the spectrum variable Alpha Canum Venaticorum the abundances of the rare earths and of elements like Cr II and Fe II vary in a period of 5.5 days with a conversion of the atoms of one group into the other. Clearly such a conversion is not physically possible. Processes must be in existence which we have not yet considered. In this connection it is significant that H. W. Babcock has discovered the existence of large magnetic fields in precisely this group of stars. In the spectrum variable HD 125248 he found a variable field whose intensity changes from approximately $+7,800$ gauss to about $-7,800$ gauss in the same period of 9.3 days which, according to Deutsch, characterizes the variation in the intensities of the spectral lines. When the Eu II lines are at their maximum strength, the magnetic field reaches its maximum value. Babcock believes that the fact that the Eu II lines are uniquely strong in this star and that the star's magnetic field at maximum is the strongest magnetic field known in nature is strong evidence in favor of the assumption that the Zeeman splitting of the lines causes a periodic variation in the measured equivalent widths, so that when the field is near its maximum value the Eu II lines are split into a very complex Zeeman pattern which, in the case of the line λ 4205, has 23 components over a range which is almost three times the width of the normal Zeeman triplet. Computation has shown that the corresponding strengthening of the total absorption of this Eu II line may easily reach a factor of about 10. This produces a change in the curve of growth, and hence a spurious anomaly in the abundance of certain atoms. There are apparently still some unexplained complications in the peculiar intensities of the spectral lines, and we must not consider this problem as having been solved. Nevertheless, Babcock's work shows in a most striking fashion that there are in existence physical processes which at times may give completely anomalous intensities to some of the observed absorption lines. Babcock has also found a strong variable magnetic field in Alpha Canum Venaticorum. He and Deutsch suggest that the presence of magnetic fields may localize the atoms of different elements in different parts of the star's surface, depending upon the magnetic susceptibilities of the different atoms.

If there were a migration of the atoms of one group, Fe II, Cr II, etc. towards one magnetic pole, and of the other group, Eu II, Gd II, etc., towards the other pole, the observations of Alpha Canum Venaticorum, HD 125248, Epsilon Ursae Majoris, and other spectrum variables might be explained by assuming that the magnetic axis makes a large angle with the axis of rotation. When the latter is approximately perpendicular to the line of sight we would observe a spectrum variable with a variable magnetic field. But when the axis of rotation is in the line of sight the spectrum would be peculiar, but constant, and the magnetic field might also be large, though constant. There is not enough material at hand to test this hypothesis. It is somewhat opposed to the theoretical prediction of a close relation between angular and magnetic moments—a relation which originally led to Babcock's discovery of stellar magnetic fields. But recent results by H. D. Babcock and G. Thiessen on the variability of the sun's general magnetic field have already undermined the appreciability of the theoretical relation to astronomical phenomena. Nevertheless, it is clear that the magnetic properties of the stars produce important changes in their line spectra and sometimes simulate abnormal abundances.

At this stage it is convenient to discuss briefly certain peculiar groups of stars which have been of considerable interest from the point of view of their chemical composition. Among them are some which possess strong emission lines of forbidden [Fe II]. These stars are rare in space. Some of them are irregular variables, while others are believed to be constant in light. WY Geminorum and Boss 1985 are striking examples (Plate VII). In the ordinary photographic region of the spectrum these stars have continuous spectra which appear to be composite. An energy distribution corresponding to a temperature of about 3000° is superposed over an energy distribution corresponding to a temperature of 10,000° or 20,000°. Similarly, the spectral absorption lines are divided into two groups. The red component which dominates the spectrum in most of the photographic region contains bands of TiO and atomic lines of Fe I, Ca I, etc. In the ultraviolet region the Balmer lines of hydrogen are prominent, as are also a few other weak lines corresponding to spectral class A or B. In addition to these features, there are strong and narrow emission lines of [Fe II]. Since the ionization potential of Fe II is 16.5 volts and the ionization potential of hydrogen is 13.5 volts, we would normally expect to see bright lines of hydrogen whenever the lines of Fe II are present. Moreover, it is not immediately clear why the permitted lines of Fe II are not also observed in emission. Ordinarily they are much stronger than the forbidden lines.

The last question was answered in an investigation by Swings and Struve, who obtained spectrograms of these stars in the extreme ultraviolet region where permitted lines of Fe II with low excitation potentials, similar to those of the forbidden lines, are present. These permitted Fe II lines were observed in emission. In the photographic region, on the other hand, the excitation potentials of the permitted lines are somewhat greater; hence the absence of these lines is probably connected with the excitation potential. In the normal sources encountered in astronomy the exciting temperatures are so high that there is no great distinction in the populations of the various levels which serve as the starting points for either the permitted or the forbidden lines of Fe II. In these peculiar stars the excitation temperatures turn out to be excessively low—one or two thousand degrees at the most—so the excitation cannot be attributed to the energy of the hotter component of each system. It might have been attributed to the cooler component were it not for a very important observation that was made some years ago at the McDonald Observatory. It so happens that one of these unusual stars is sufficiently close to us to be actually resolved in the telescope as a visual double star. This system is Antares, or Alpha Scorpii. It consists of a first-magnitude red supergiant and a sixth-magnitude blue under-luminous helium star. The separation between the two stellar components is 3″. Ten years ago at the Mount Wilson Observatory, O. C. Wilson and R. F. Sanford discovered that the spectrum of the faint blue companion of Antares contains emission lines of forbidden [Fe II].

Under excellent atmospheric conditions the spectrum of the binary can be completely resolved. On one occasion the light of the B-type component was held precisely in the same place on the slit of the spectrograph, without permitting the image to trail back and forth along the slit. The result was an overexposed spectrum of the B-type star with the characteristic diffuse and faint absorption features of H and He I. The emission lines of [Fe II] stood out as tiny black arrows on both sides of the band of continuous stellar radiation. These small black arrows correspond to radiation from regions outside the B-type star. Consequently the source of the forbidden lines is not the atmosphere of the B-type star, but a small approximately round nebulosity centered upon it. This nebulosity is the medium which produces bright lines of Fe II, both forbidden and permitted, having an excitation temperature much too low to be identified with the B-type star, but having no emission lines of hydrogen. Clearly, the low excitation temperature can best be explained through the process of collisions. The forbidden levels are sufficiently low to allow collisions to play an important role if other factors are favorable. The absence of hydrogen, on the other hand, is definitely abnormal because in all ordinary emitting nebulae hydrogen is by far the most prominent feature, and in the planetary and diffuse galactic nebulae the abundance of hydrogen is quite similar to that observed in the normal sequence of stars. It is probably necessary to suppose that hydrogen is deficient in the nebulosity surrounding the B-type component of Antares. An attractive idea is to suppose that this nebulosity is not completely gaseous, but that it consists of solid, perhaps even meteoric, particles, which have long ago lost most of their hydrogen and are rich in iron. We certainly possess such a nebulosity in the solar system, where it leads to the phenomenon of the solar corona on the one side and to that of the zodiacal light on the other. It is not unreasonable to suppose that meteoric particles could become vaporized in the vicinity of the B-type component and that emission lines of Fe II would be observed. This process would be analogous to that which gives rise within the solar system to the characteristic emissions of the comets. Although the latter, when at great distances from the sun, show only the lines of molecules composed of the lighter gases, there have been observed strong emission lines of Na I in comets which have approached the sun to a relatively small distance. Very roughly, the diameter of the nebulosity around the B-type component of Alpha Scorpii is about ten times larger than the solar system. Despite the fact that from a representative point within the nebula the red star appears as a disk which is about 200 times larger than the blue star, it is the light of the latter in the extreme ultraviolet region which is responsible for the ionization of the gases, though not necessarily for the excitation of the low-lying atomic levels.

As we approach the region of the cooler stars in the H-R diagram we again find a number of special groups of objects which present interesting anomalies of chemical composition. There is, first of all, a remarkable difference in the relative intensities of bands produced by the two carbon isotopes C^{12} and C^{13}. On the earth the ratio is about $C^{12}/C^{13} = 90$. In the N-type stars A. McKellar and G. Herzberg found a much greater relative abundance of C^{13}. In the R-type stars McKellar found two groups. In three stars the ratio C^{12}/C^{13} was greater than 50—which could very well mean that it is identical with the ratio observed on the surface of the earth. But in fifteen other stars the ratio turned out to be somewhere between 3 and 4. McKellar reached the conclusion that it is not possible to explain this difference as the result of differences in the curves of growth of the two groups of objects. He therefore felt that the difference in composition is real. Two suggestions have been advanced to explain this difference. The one is based upon the work of O. Klein, G. Beskow, and L. Treffenberg, who had found that in closely packed cores of prestellar material a ratio between 10 and 50 might have been produced in thermal equilibrium at about 1,000,000 electron volts. The other suggestion, advanced by E. Fermi, presupposes that in the original medium of the stellar substance the ratio C^{12}/C^{13} was about 3. But the capture cross section of the nuclei of C^{13} for protons is about 70 times larger than that of C^{12}. Hence, if we start out with an abnormally large fraction of C^{13}, we would rapidly convert it, through successive proton captures, first into N^{14} and then into O^{15}; the latter rapidly disintegrates into N^{15} and finally into C^{12} and He^4. The consequence of this process is to convert quite rapidly the excessive amount of C^{13} into C^{12}, so that equilibrium is reached with a ratio of C^{12} to C^{13} in the neighborhood of 70.

This would mean that the Bethe cycle must have been operating for a long time, and the question to which we as yet have no answer is whether the effectively complete conversion to a ratio of even more than 70 in the sun, the earth, and in many stars would not at the same time require a far-advanced state of conversion of hydrogen into helium. In the case of the interstellar gas we only know from the work of O. C. Wilson that the ratio must be larger than about 5. Consequently, it is possible that the interstellar medium may have a composition that differs only slightly from the ratio of 3, which on Fermi's hypothesis corresponds to the original condition.

We have no assurance, however, that the present composition of the interstellar gas really resembles that of the original medium. It is possible that there is a continuous exchange of matter between stars and gaseous substratum.

This problem is one of very great interest, and if it stood alone it would provide a strong confirmation of the operation of the Bethe process; but since in other respects we have found only contradictory evidence and evidence which seemed to suggest that there is no complete mixing of the different regions in stars, it is perhaps advisable at this stage to await further developments. It may be possible to explain the divergent abundances of the carbon isotopes in terms of other physical processes.

Still another interesting anomaly which was discovered by McKellar in certain late-type stars is the great intensity of the Li lines in some variables, especially WZ Cassiopeiae. In a study of more than 40 red carbon stars McKellar and W. H. Stilwell found that there is no marked correlation between the equivalent widths of the Na I and Li I resonance lines or between the spectral types of the stars, as classified by other criteria, and the intensities of these two lines. The authors remarked that Na I and Li I arise from similar transitions and that the ionization potentials are almost the same, 5.13 and 5.38 volts, respectively. Hence, they concluded that if the relative abundance of Na to Li were constant, there should exist a strong correlation between the two intensities. The absence of such a correlation might indicate peculiar differences in the abundances of the two elements. But this is not certain, and the authors themselves pointed out that according to the work of Keenan and Morgan one might expect a strengthening of the Li I lines in such stars as WZ Cassiopeiae, simply as the result of very low excitation temperatures. We have again not much more than a suggestion of an interesting possible difference in the abundances of some of the elements. If it could be proved that Li is really more abundant by a factor of perhaps 100 in WZ Cassiopeiae than in the sun, the consequence would be in many respects remarkable. Swings has stressed the importance of differences in continuous absorption which may in effect reduce the depth in the atmosphere to which we can see in one part of the spectrum as compared to another part. Hence, caution must be exercised in all studies of stellar spectra that are rich in molecular bands. Spitzer has attempted to determine whether the abundance of Li, Be, and B is higher in the interstellar gas than in the atmosphere of the sun. His observations were negative, but he felt justified in suggesting that the low terrestrial abundance of these elements, especially of Li and Be, probably extends also to the interstellar material. Li must have become exhausted before most of the stars were formed. The only possible explanation of the high abundance in such a star as WZ Cassiopeiae would be to suppose that it was formed at a very early stage in the life of the galaxy when Li was more abundant than it is now. The depletion of Li in the interstellar material observed at the present time may be the result of a continuous exchange between the stars and the interstellar clouds.

In the late-type stars, interesting conclusions have been obtained independently by L. Rosenfeld and by H. N. Russell from the intensities of certain molecular bands. For stars which are similar in composition to the sun, in which oxygen predominates over carbon, the oxides are strong. For example, TiO is the most characteristic feature of the M-type stars. In the S-type stars ZrO is unusually strong, so that Zr may have more than its normal abundance. In the so-called carbon stars of spectral classes R and N, the cyanogen bands, as well as the bands of C_2 and CH, are relatively much stronger than in the stars of classes M and S. This interpretation in terms of differences in the abundances of oxygen and carbon was suggested many years ago by R. H. Curtiss. It was confirmed by means of accurate computations of the equilibria of the principal molecules by Russell. As an example, we may consider the molecule TiO. The number of TiO molecules per unit area reaches a maximum of 10^6 in the giant oxygen stars and of 10^5 in dwarf oxygen stars. In giant carbon stars the corresponding values are 10^1 for the giants as well as for the dwarfs. In these computations the usual abundances of carbon and oxygen were interchanged. If we next consider the molecule CH, we find that in the giant oxygen star it reaches a value of 10^4 and approximately the same for the dwarf oxygen stars. In the carbon stars, on the other hand, the maximum value of CH is approximately 10^8. These results are suggestive, but at the same time they are disquieting in view of the fact that we have discovered no striking differences in the abundances of the corresponding atoms, C, O, and N among the B-type stars, where we observe only the atomic lines.

There remains the great mass of F, G, and K stars. Superficially the spectra of these stars, on the main sequence and among the giants, display few spectroscopic anomalies which could be explained in terms of differences in abundance. But this is partly caused by the complexity of the spectra. There are indications, especially in the work of Bidelman, that more detailed studies will bring to light a diversity of spectroscopic features analogous to that already found among the A-type stars. But at present, these differences do not necessarily suggest that the abundances are also appreciably different.

12. Summary

Our principal result is the fact that the atmospheres of the vast majority of stars of all kinds—dwarfs, giants, supergiants, and even white dwarfs—have approximately similar chemical compositions. This composition is similar to that of the interstellar gas, the bright diffuse nebulae, and the planetary nebulae. It differs from that of the meteors, the comets, and the terrestrial planets, in the abundances of the light gases.

There are, however, some stars which show anomalous spectra. Among these are many objects classified as members of Population II, which probably were formed in the region of the galactic center. Some of the peculiar line-intensities can be explained only in terms of different abundances of certain elements. But these differences do not present any discernible features which would permit us to attribute them to the operation of a single physical process such as the Bethe-v. Weizsaecker cycle of transformation of H into He. In a few cases we suspect the operation of selective radiation pressure which may have driven away atoms of certain kinds, leaving behind a residue with anomalous composition. Perhaps this same process may have built up outer shells of the opposite composition.

There are a few indications that stars which, from their behavior in the L, $\mathfrak{M}$, R relations, are believed to be deficient in internal H content, show spectroscopic features of higher than normal luminosity.

The H-R diagrams for galactic clusters present the best evidence we now possess in support of the Bethe-v. Weizsaecker process. They also give strong observational support to the conclusion that the stars of types O, B, and A evolve fairly rapidly and are somehow completely removed from the upper left-hand portion of the H-R diagram. The spectroscopic anomalies at the left-hand sides of cluster diagrams suggest that internal processes are at work, but the spectra themselves cannot be explained with confidence as the result of H deficiency in the atmosphere. There is probably no complete mixing.

The H-R diagrams of the clusters also suggest that the disappearance of the O, B, and A stars in the older groups is not accompanied by the appearance of an equal number of red giants or other unusual objects having approximately the same mass as the normal blue stars. Instead we must probably think of two simultaneous processes of evolution—one which removes a hot star from the main sequence and moves it over into the region of the red giants, and another which permits it to slide along the main sequence, with a corresponding loss of mass, and which produces an accumulation of red dwarfs.

We shall see that there are strong observational and theoretical grounds for believing that both processes are really at work.

II

SOME PROBLEMS OF STELLAR EVOLUTION

1. The Energy Content of the Stars

THE amount of radiant energy which is produced by the sun and emitted into space amounts to about 10^{41} ergs per year. At the present time the sun contains, within its mass, radiant energy amounting to about 3×10^{47} ergs, translational energy of the atoms and electrons amounting to about 3×10^{48} ergs, and energy of ionization and excitation amounting to about 3×10^{48} ergs. All these immediately available stores of energy would suffice to cover the sun's annual expenditure of energy over a period of only about fifty million years, and this is entirely insufficient to account for the fact that the radiation of the sun has remained constant over an interval of time which, from geological data, must have been many millions of years. We have convincing evidence that the radiation of the sun was approximately the same as it is now, several hundred million years ago, when the fossil layers of coal were formed.

The measurements of the lead and the helium contents in the rocks of the earth can be used to determine the age of the solid crust of the earth if we assume that all observable traces of these elements have been formed within the rocks as the result of the transformation of uranium and thorium by means of radioactive processes. In this manner the age of the earth's crust has been estimated at about 2×10^9 years. In a similar manner the ages of the meteorites may be estimated at about 10^9 years for the average of several samples which show a considerable amount of scatter.

It is reasonable to suppose that the radiation of the sun has not changed greatly during this interval of about 2×10^9 years since the solid surface of the earth was formed. Now, if we multiply the annual expenditure of radiant energy by this number of years, we find that the sun has lost more than 10^{50} ergs. Remembering that the total mass of the sun is 2×10^{33} grams, we must conclude that on the average every gram of substance within the sun has produced at least 10^{17} ergs. This is a tremendous amount of energy; it may be compared to the amount of heat which is developed when a gram of coal is completely burned, namely 3×10^{11} ergs. Hence, we see that the sun contains material which is about a million times more efficient in

PLATE VIII. Dark clouds in the constellation Taurus. (Photograph by E. E. Barnard.)
The emission-line star in B10 is at the center of the round nebulosity, about ½ inch to
the right of the plate-center.

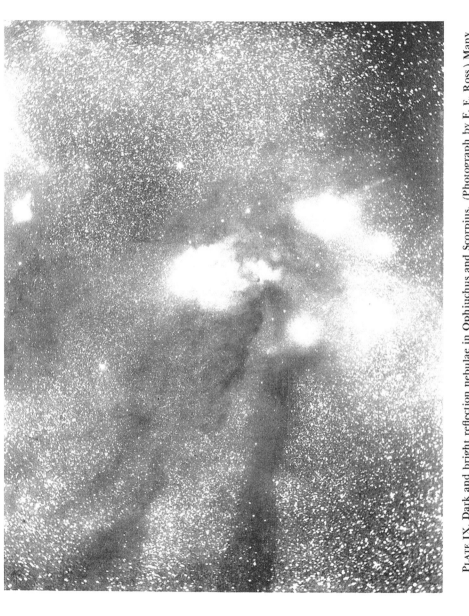

PLATE IX. Dark and bright reflection nebulae in Ophiuchus and Scorpius. (Photograph by F. E. Ross.) Many of the faintest stars immersed in the dark clouds have strong emission lines. The central stars of the luminous reflection nebulae are of spectral class B_2 to $B8$.

producing energy than is the most efficient chemical process known on the earth.

For more than a century astronomers have searched for the source of this tremendous power of energy generation. About one hundred years ago J. R. Mayer developed a theory in which he attributed the light and heat of the sun to the impact of meteorites. This hypothesis was later developed by Norman Lockyer, who attempted to explain the physical properties of the stars and their spectra. A recent revival of the meteoric hypothesis, by F. Hoyle and R. A. Lyttleton, resulted in the current ideas concerning the growth of stellar masses in cosmic dust clouds. Although it is not now believed that this source of energy is sufficient to maintain the production of radiant energy of a star over long periods of time, the idea of mass accretion has been particularly fruitful in explaining the origin of the stars and their possible growth from the stage of small condensations in interstellar matter to the stage of protostars, or stars of very diffuse constitution and high surface temperature.

From the observations of meteors it is known that they strike the earth or its atmosphere with an average velocity of about 40 km/sec. The amount of heat which is generated per gram of meteoric substance is therefore approximately 200 calories, this amount of heat representing the equivalent of the kinetic energy of the meteor.

A meteor striking the surface of the sun would have a velocity about fifteen times greater than the velocity of impact at the surface of the earth, or about 600 km/sec. Since the heat generated in the process of impact is proportional to the square of the velocity, or in this case 225 times the value found for the earth, we find that the amount of heat generated by one gram of meteoric material at the surface of the sun is about 45,000 calories. If the sun's heat were entirely accounted for by the impact of meteors, then the amount of heat received by the earth from the sun would represent that fraction of the total amount created by all the meteors falling upon the sun, which the surface subtended by the earth at the sun is to the surface of the entire sphere, whose radius is one astronomical unit. The earth intercepts this same fraction of meteors out of the total, provided that we make the reasonable assumption that they fall into the sun from very great distances. Hence, the earth would receive heat from the impact of this small fraction of meteors, but the amount of this heat would be only 1/225 of that which the earth receives from the sun. In reality, however, the eight million or more meteors which strike the earth daily provide only an infinitesimal fraction of the computed amount of heat on the earth. Consequently it is quite certain that at the present time the impacts of meteors are not sufficient to maintain the output of radiant energy of the sun.

In 1854, H. v. Helmholtz attempted to show that the contraction of a gaseous sphere by its own gravitation would provide a source of energy that might account for the heat of the sun. This process, though far more efficient than that of meteoric impacts, is also insufficient. A shrinkage in the sun's diameter of about 1/4,000 of its present value would, according to F. R. Moulton, maintain its present radiation over an interval of 10,000 years. If the sun contracted its radius by 37 meters per year, this would suffice to maintain the present rate of radiation. Such a contraction could not now be observed, but over an interval of 2 billion years it would result in the complete disappearance of the sun, so we must seek for other processes which are much more powerful than even the process of contraction. This conclusion can be strengthened if we compute the total amount of gravitational energy which would have been produced if the sun had contracted from infinity to its present radius of 7×10^{10} cm. Such a contraction would have produced a total of 6×10^{48} ergs, but only a fraction of this was available for heat radiation. This is much less than the energy which has been radiated into space during the lifetime of the earth.

Eddington has shown that the gravitational energy obtained from the contraction of the sun from infinity to its present radius would have been sufficient to maintain its present output for 23 million years. In reality, the luminosity of the sun was probably less when the radius was larger. If allowance is made for this change in luminosity, the interval of time would be doubled.

On the other hand, if we start with the sun at a temperature of 3,000° and compute how long it would have taken by contraction to maintain its radiation at the present rate, the result would have been approximately 15 million years, which, as we have seen, is a short interval of time compared with the age of the earth.

Eddington has pointed out that the conclusion would be even more surprising if we had taken a giant star rather than the sun. For example, he has computed that a star whose mass is about ten times that of the sun would require only 30,000 years to change from a temperature of 3,000° to one of 6,000°, or 70,000 years from a temperature of 3,000° to one of 10,000°. That such a rapid evolutionary process does not exist among the giants is shown by the variable stars of the Delta Cephei type. Their periodic changes in brightness are the result of pulsations in volume which can be shown to obey the relation

$$P\sqrt{\rho} = \text{constant.}$$

If the variable star is contracting, the density ρ changes, and consequently the constancy of the product $P\sqrt{\rho}$ requires that the period

P must also change. According to Eddington, if the energy generation were to be attributed to contraction this change would amount
to 17 sec/year. Actually no such change has been observed. There
are small changes in period, but some stars have increasing periods
while others have decreasing periods. In the case of Delta Cephei
the observations indicate that the period may be variable, with a rate
of change that is about 150 times slower than that predicted by the
contraction theory. Hence the lifetime of a star such as the sun, instead of being approximately 40 million years, would have to be
about 6 billion years; and this is ample for all requirements.

Although the theory of contraction is not sufficient to account for
the age of the sun, it undoubtedly has an important bearing upon the
problem of stellar evolution. A star which has been recently formed
and has not developed a high internal temperature undoubtedly
contracts and generates heat in the manner proposed by Helmholtz.

2. The Transformation of H into He.

If we accept three billion years as the approximate age of the universe, the only process of energy generation which is adequate to give
the required amount of heat and light is the conversion of hydrogen
into helium according to the formula

$$4H \rightarrow He + Energy.$$

In this expression mass is lost because the weight of four hydrogen
atoms is $4 \times 1.008131 = 4.032524$, while the weight of one He atom
is 4.003860. The difference is 0.028664, and this amount of mass is
converted into radiant energy by the Einstein equation

$$E = mc^2,$$

where m is the mass converted in the process and c is the velocity of
light. Since in every case this small amount of mass is lost whenever
a helium atom is created out of four hydrogen atoms, the fraction of
the mass that is converted into energy is

$$\frac{0.028664}{4.003860} \sim 0.007.$$

We see at once that the loss of mass in this process is extremely
small. Hence, a star which evolves in this manner must move along
a line or surface of virtually constant mass. In this connection it is
necessary to be cautious. Usually we draw in the H-R diagrams a set
of curves or approximately straight lines corresponding to equal
masses which result from the mass-luminosity relation. But this relation is correct only when the chemical composition of a star remains

the same. In the process of conversion of H into He the chemical composition does not remain constant and the curves of constant mass will therefore be different from the ones which rest upon the application of the mass-luminosity curve.

In the process of evolution the internal constitution of a star must change, and we cannot be certain on theoretical grounds what form the mass-luminosity relation will have. But we can perhaps obtain some information from the observations. Unfortunately we know very little about the masses of the supergiant stars. As far as data are available, these stars conform to the empirical curve in Figure 1. Thus, V. Goedicke has obtained the masses and luminosities of the components of the eclipsing binary VV Cephei, which consists of an M-type supergiant and a normal B-type companion. The solution of the orbit is difficult, and Goedicke made two assumptions: (a) that the M star has a sharp photosphere, and (b) that its edge is diffuse and the light of the B star shines through it, as is the case in Epsilon Aurigae in which the light of the F component shines through the semitransparent outer parts of the supergiant infrared component. With these assumptions, Goedicke finds:

	(a) Sharp photosphere	(b) Diffuse photosphere
$\mathfrak{M}_M$	$57\mathfrak{M}_\odot$	$47\mathfrak{M}_\odot$
$\mathfrak{M}_B$	$40\mathfrak{M}_\odot$	$33\mathfrak{M}_\odot$
$M_M(\text{bol})$	-9.3	-7.4
$M_B(\text{bol})$	-5.2	intermediate
R_M	$3000R_\odot$	$1220R_\odot$
R_B	$22R_\odot$	intermediate

Both sets of values ($\mathfrak{M}_M$, M_M) lie, within the possible limits of error, on the usual mass-luminosity curve. This conclusion is confirmed by other similar systems. Hence we can regard the lines of equal mass in Figure 8 as having been observationally confirmed. A star of early spectral type whose evolution proceeds without loss of mass probably moves along one of these lines. But there remains the possibility that although the general trend of the evolutionary paths is that indicated by these lines, there may be small irregularities which are not consistent with the mass-luminosity relation.

For stars built on the standard model E. Schatzman found from theoretical considerations that in the course of converting H into He such a star would move upward, in the H-R diagram, because the luminosity increases with decreasing H content. At the same time these stars would move to the left remaining either slightly above or slightly below the main sequence. If real stars behaved in this manner we should now find a large dispersion of masses for any given early spectral type. The less massive stars would be the ones which had converted a large fraction of their original supply of H into He. It is probable that the O, B and A-type binaries actually have a fairly large dispersion of $\mathfrak{M}$. For example, the B2-type spectroscopic binary VV Orionis is believed to have masses of the order of $\mathfrak{M}_1 = \mathfrak{M}_2 = 1.7\mathfrak{M}_\odot$, while according to J. A. Pearce the B2-type binary HD 227696 has $\mathfrak{M}_1 = 16.7\mathfrak{M}_\odot$ and $\mathfrak{M}_2 = 12.8\mathfrak{M}_\odot$. However, as far as we know, these systems all define a narrow mass-luminosity relation. Moreover, the evidence of the galactic clusters unmistakably points toward an evolutionary process of the kind described in the preceding chapter. It would be unsafe to accept the observational evidence as being contrary to the path described by Schatzman. But we may at least tentatively search for a theoretical cause—perhaps in terms of the internal constitution of the stars—that would permit us to reconcile the observations of the clusters with the results of Schatzman.

But let us now estimate how long the sun could continue radiating at its present rate if the entire mass of 2×10^{33} grams consisted of H. We have just seen that only 0.007 of the mass can be converted into energy; hence the number of grams which can be so converted will be 1.4×10^{31}. The number of ergs which we can realize in this process is obtained by multiplying this mass by the square of the velocity of light, or 9×10^{20}. The result is approximately 10^{52} ergs. But the sun radiates energy at the rate of 10^{41} ergs per year. Hence, dividing one number by the other, we obtain 10^{11} years as the approximate length of time that the sun could radiate at the present rate if its entire mass consisted of hydrogen.

In reality, only about 40 per cent consists of hydrogen. Thus the interval of time the sun could continue radiating at its present rate would be about 4×10^{10} years. P. Ten Bruggencate has computed this quantity more accurately, making allowance for the fact that, as the process of conversion of hydrogen into helium goes on, the luminosity of the sun increases and the rate of energy generation becomes greater and greater. The quantity we have derived is therefore an upper limit. On the other hand, we must remember that the present helium content of the sun, amounting to perhaps 10 per cent by numbers of atoms, may already represent a substantial increase over the original composition. Extrapolating the life history of the sun into the past, we might conclude that its present age may well be of the order of 2 or 3 billion years.

In the same way, we can compute the approximate ages of the other stars in the H-R diagram. For each star the luminosity measures directly the amount of energy which is produced by the thermonuclear process. Therefore the lifetime of a star is inversely proportional to its intrinsic luminosity. Similarly, the age of a star is directly proportional to the amount of convertible material, and this is given by its mass. If we know the luminosity and the mass of a star, we obtain immediately the corresponding age from the relation

$$T_*/T_\odot = \left(\frac{\mathfrak{M}_*}{\mathfrak{M}_\odot}\right)\left(\frac{L_\odot}{L_*}\right),$$

where T_* designates the age of the star, $T_\odot$ the age of the sun, $\mathfrak{M}$ the mass, and L the luminosity. In this manner we obtain the following values of T_*, which have been taken from a more detailed computation:

Sp	Temp.	$L_*/L_\odot$	$\mathfrak{M}_*/\mathfrak{M}_\odot$	T_*
O	30,000	10^5	14	10^7
B	20,000	10^3	10	$10^{8.5}$
A	10,000	10	2.5	$10^{9.7}$
F	8,000	2.5	1.3	$10^{10.3}$
G	6,000	1	1.0	$10^{10.6}$

The maximum age of an O-type star is only 10 million years—a remarkably short interval for a massive star like 10 Lacertae or Zeta Puppis. Since the values of T_* are only upper limits, the real ages of such stars may be even less. Moreover, there are in existence stars whose luminosity is greater than that of a normal O-type star. For example, a bright supergiant may have a luminosity almost a million times greater than that of the sun. With a mass not much in excess of 30 times that of the sun such a star can have existed in its present form not much longer than about 1,000,000 years. This amazingly short age is perhaps the most significant result which comes out of the theory of nuclear processes in the stellar interiors. Astronomically speaking, these ages are so short that we can readily believe that the stars are being formed, even at the present time. Our problem is to try to interpret correctly what we actually observe in the sky. The process of star formation must be going on before our very eyes, and if we have not discovered it thus far our failure must be due to the fact that we have overlooked some simple phenomenon in the universe.

The average velocity of an O-type star with respect to the neighborhood of the sun is about 10 km/sec. In an interval of 10^7 years, such a star would have traveled a distance of about 3×10^{15} km or about 100 parsecs. Remembering that the distance of the galactic center from the sun is 10,000 parsecs, we immediately see that during its entire existence an O-type star cannot have moved very far away from the neighborhood in which it was born. Hence, if we speak of the region surrounding the solar system in our galaxy as the local swimming hole, we must conclude that all the early-type stars within that swimming hole were actually produced in it and our present photographs of the Milky Way must show us the very regions in which they were born. The only possibility is that they were formed out of the clouds of interstellar dust and gas which occur in our local swimming hole at distances of several hundred parsecs from the sun. These clouds of obscuring matter are often seen projected in front of dense star clouds of the Milky Way, many of which are very much farther away (Plate I). In the present state of our galaxy, the amount of mass in the dark clouds is approximately equal to that concentrated in the stars. Consequently there is even now an ample supply of diffuse matter which may form the substance out of which future stars can be condensed.

C. F. von Weizsäcker has recently proposed the idea that the stars may all have been formed originally billions of years ago, but that the hotter stars have experienced in relatively recent times a process of "rejuvenation" through the accretion of interstellar matter. In principle, this idea does not differ much from the one we have discussed because the process of accretion may well be considered as a delayed mechanism of star formation.

3. The Formation of Stars in Interstellar Clouds

The problem of the origin of a star through the process of condensation of diffuse matter has been discussed by L. Spitzer, Jr., F. L. Whipple, and B. J. Bok. The latter has called attention to the existence in the dark clouds of small spherical globules which have diameters ranging from 0.06 parsec to 0.5 parsec and are gravitationally stable in the sense that they are able to retain their mass through the action of their own gravitation. Bok has computed that a typical globule may have a mass equal to 2.5 times the mass of the sun. Very small globules have masses of the order of 0.1 that of the sun, and the largest of all known globules, namely the great southern "Coal Sack," has a mass about 650 times that of the sun. A globule of interstellar matter, once it has been formed, will collect dust and gas from the outside and will therefore gain in mass. The smallest globules increase quite slowly, but one of average mass approximately doubles in 3×10^8 years, and a large globule like the Coal Sack would become twice as massive in 3×10^7 years. The exact process of star formation is not known, but theoretical considerations have led Whipple to suggest that interstellar clouds without rotation but with turbulent viscous friction will collapse in approximately 10^8 years, because the pressure produced by the turbulent motions of the original cloud will rapidly decrease as the clouds cancel out their motions.

All indications are that interstellar matter can and does condense into protostars which later form real stars, but we do not definitely know whether the newly created stars appear at the upper end or at the lower end of the main sequence in the H-R diagram.

At this point it may be useful to recall that by far the greatest number of stars in our galaxy belong to the main sequence. We have already seen (p. 43) from the computations of Parenago that the number of stars in the main sequence is approximately 10^{11}. The next most numerous group of objects in the H-R diagram is that of the white dwarfs. There are perhaps 10^9 or 10^{10} such objects in our Milky Way. The number of giants and supergiants is probably less than 10^8. That of the subdwarfs is not yet known, but it may be not too different from that of the ordinary main sequence. The reason for this is that the subdwarfs belong to Population II. Since the O and B stars are all relatively young objects, it is natural to suppose that they have been formed recently out of the interstellar matter; but this does not preclude the formation of main-sequence dwarfs out of the same material.

It is at present somewhat disconcerting that we cannot tell definitely whether a dwarf is or is not formed directly out of the interstellar matter. Ambarzumian has recently discussed the cosmogonical consequences of the H-R diagram and has called attention to the fact that stellar aggregations, such as galactic clusters, often contain the entire main sequence from spectral class O to M. For example, the cluster NGC 6231 contains several emission-line O-type stars, a few P Cygni-type stars and two Wolf-Rayet stars. In addition, there is the usual main sequence down to, at least, spectral classes F and G. The peculiar stars, for example, those of the P Cygni-type, can hardly be older than 10^7 years. During this interval of time the main-sequence dwarfs of types F and G would have undergone hardly any change as the result of nuclear transformations. Consequently, Ambarzumian was led to believe that the arrangement of the cluster in the H-R diagram was probably from the beginning what it is now. In other words, the stars were all formed exactly as they are at the present time. He does not believe that they had evolved appreciably by sliding down the main sequence.

In exactly the same way, the double cluster of Perseus would also compel us to believe that the stars within its boundaries were formed as B-type, A-type or F-type main-sequence stars, or even as red supergiants, and were not appreciably changed during the relatively short lifetime of this grouping. Ambarzumian's reason for drawing this conclusion is the fact that along the main sequence the mass changes from a value of about 30 times that of the sun, near spectral class O, to a value of the order of one-tenth the mass of the sun, at the lowest observable temperatures. Hence, evolution along the main sequence must be accomplished by a change in mass, and such a change cannot be produced by the carbon cycle of nuclear transformations or, for that matter, any other known process of nuclear transformation. The complete annihilation of material which was believed to play a role when astronomers first began to think in terms of Einstein's equation for the equivalence of mass and energy has been abandoned because it seems to contradict well established principles of physics.

Several large obscuring clouds in the Milky Way have been investigated with respect to their distances and opacities by comparing star counts made in the area covered by the cloud and in transparent regions of the Milky Way. Since the nearest dark clouds are approximately 100 parsecs away, there are numerous faint stars which are seen as though projected on the dark cloud, but they are really closer to us. In the densest parts of the dark clouds the opacity is so great that no light from distant parts of the Milky Way can penetrate, but there are some stars which are embedded within the clouds themselves. If we know the spectral type of a star belonging to the main sequence, we can estimate its distance with a considerable degree of confidence because, as we have seen, the spread of the main sequence in magnitude is not very large. For example, if the spectrum of a star is found to be M0, then from the H-R diagram we immediately read off that the absolute bolometric magnitude is $+9.5$. If the nebula is at a distance $D = 100$ parsecs, we obtain for the apparent magnitude $m = 5 \log D - 5 + M = 10 - 5 + 9.5 = +14.5$. This would be the apparent bolometric magnitude of a star at the distance of the nebula, but not obscured by it. If we had found photometrically an apparent bolometric magnitude of $+16$ or $+17$, we would have been justified in concluding that the star is involved in the nebulosity and experiences an absorption of about two magnitudes in it. It must be remembered that the bolometric magnitude represents a measure of the total amount of radiant energy in all frequencies, and not only that which is transmitted by the atmosphere of the earth. In reality we observe only a fraction of the light because air is not transparent to all wave lengths. Hence the visual magnitude is obtained from the bolometric magnitude by applying to the latter a correction designated as the bolometric correction. If we call the latter Δm, we obtain

$$m_{\mathrm{vis}} = m_{\mathrm{bol}} + \Delta m.$$

In the case of a star of spectral type M, the bolometric correction is quite large: $\Delta m = 2$ magnitudes. Hence, we should have expected the star to appear, not with a visual magnitude of 14.5, but with one of 16.5.

In this manner it is possible to distinguish, with a reasonable degree of certainty, whether a star is or is not involved in a nebulosity.

In 1946, A. H. Joy found that the faint stars which are involved in the obscuring nebulosity of the constellation Taurus (Plate VIII) possess spectra that are different from those in the open regions of the Milky Way. The usual appearance of a stellar spectrum is that of a continuous band of light upon which are superposed numerous absorption lines. Emission lines are rare, especially among the dwarf stars of spectral types G, K, and early M. Joy first noticed that the so-called T Tauri variables are always associated with wisps of dark and luminous nebulosity, and that these stars are at the same time characterized by strong emission lines in their spectra. This suggested to him the possibility that all stars of the main sequence and of low temperature may well be characterized by emission-line spectra of the T Tauri type. Objective-prism spectrograms were obtained at the Mount Wilson Observatory and promptly led to the discovery of some forty stars in the dark clouds of Taurus which have strong emission lines of hydrogen, ionized calcium, and occasionally also of ionized iron, neutral helium, and a few other elements. Greenstein next computed the expected numbers of stars involved in the dark clouds of Taurus. He obtained this number by means of the so-called luminosity function, which gives the number of stars per cubic parsec, in the vicinity of the sun, for every interval of one magnitude in absolute luminosity. The volume of space in the dark nebulae of Taurus investigated by Joy is about 780 cubic parsecs. From the luminosity function Greenstein computed that there would be about two stars per hundred cubic parsecs. This would give a total of seventeen stars within the nebula. The actual number observed by Joy is almost two and one-half times greater. Few of Joy's emission-line stars can be ordinary foreground dwarf Me stars of the kind we occasionally find even in the open regions of the Milky Way, because, according to Greenstein, the total number of the foreground stars is only about ten. It is reasonable to conclude not only that every dwarf star embedded in the nebula possesses emission lines, but also that the density per unit volume of these faint peculiar dwarf stars is greater than in the open regions of the Milky Way.

More recently, M. Rudkjøbing and O. Struve investigated, in a similar manner, the dark region in the constellations Ophiuchus and Scorpius (Plate IX). The work was done entirely by means of a slit spectrograph, since the stars were all very faint and would have been blended with the light of the night sky unless the latter were removed by means of the slit. The total number of stars investigated was twenty-six. Of these, six have strong emission lines of hydrogen, ionized calcium, and in a few cases He I and Fe II. Most of the non-emission stars are either foreground objects or stars which are shining through the thinner parts of the nebula and are therefore farther away than the cloud. Seventeen additional emission-line dwarfs in the same region were discovered in 1949 by G. Haro with the Tonan-zintla Schmidt telescope in Mexico. It is reasonable to suppose that in the case of this nebula, also, practically all the faint red dwarfs of spectral classes K and M show strong emission features. One of these stars was found to be variable in light. It may turn out to be a real T Tauri variable.

The dark and bright nebulosities associated with the Orion nebula contain many irregular variable stars. The brightest of these are of spectral class A and contain no emission lines. Greenstein and Struve found only one K-type star with emission lines among these brighter variables. Reaching a lower apparent luminosity G. Herbig found a larger fraction of T Tauri type emission line objects. Finally, among stars of apparent magnitudes 14, 15 and 16, Haro found more than 100 of these peculiar spectra. He has estimated that in a comparable region of the Milky Way which is free of nebulosity the number of emission-line stars would have been not in excess of 10.

There are strong indications that not all emission-line stars associated with dark nebulae are similar in character. The typical T Tauri variables have much more complex spectra than the majority of the faint emission-line dwarfs in Taurus or in the clouds of Scorpius and Ophiuchus. Among these dwarfs there appear to be several different kinds. A few are blue in color, with their hydrogen emission lines extending far out into the ultraviolet region of the spectrum, toward the Balmer limit. Such a star is the central object of a small luminous patch within the large dark cloud of Taurus which Barnard designated many years ago as object B10 in his catalogue of dark objects in the sky (Plate X). The nebulosity itself is apparently of the reflection type. There is no trace in its spectrum of any emission of forbidden [O II], which is usually strongly present when the nebula emits light. But the size of the luminous nebula is much too great to be explained in terms of pure scattering or reflection of light by white particles. If the absorption of the star's light is similar to that observed for an average reflection nebula, we must conclude that this object is about seven or eight magnitudes fainter than it would have to be to produce, by pure diffuse reflection, a luminous nebula with a radius of five minutes of arc. But the continuous spectrum of the star is strong, even at the limit of the Balmer series and beyond it. This suggests that interstellar reddening cannot be great. Undoubtedly the star is really quite faint; its apparent magnitude is about 15, and there is no reason to suppose that the photographic absorption is more than one or two magnitudes. The Taurus nebulosity is at a distance of roughly 100 parsecs. Hence, at a distance of 10 parsecs the photographic magnitude of the central star would be about 10, uncorrected for absorption, or perhaps 8 or 9 after correction. This would be its absolute magnitude, and the corresponding spectral class would be between K5 and M0. But the continuous spectrum between the emission lines is certainly not that of an ordinary K or M star. It is almost continuous and it contains much more blue light than is consistent with so late a spectral class. We must conclude that this star, and other objects similar to it, though in absolute magnitude they resemble main-sequence objects of class K or M, are nevertheless peculiar in the sense that they are characterized by a continuous emission that corresponds to a much hotter temperature.

Such a temperature is probably not physically real. At least we do not at present believe that we are concerned with a star of the white-dwarf class, because Joy has pointed out that there are other objects which show a similar film of continuous emission on top of a spectrum displaying traces of the usual late-type absorption features. The majority of the stars in his list, as well as those in Scorpius and Ophiuchus, are clearly dwarf stars with absorption features of classes K and M. It is only in some of these objects that there is an additional source of continuous emission which obscures the absorption lines and tends to make the star look bluer than a normal red dwarf.

The discovery by Luyten of a dwarf star which is subject to irregular outbursts or "flares" in luminosity with simultaneous changes in spectrum suggesting the superposition of B-type features in absorption and emission and of a blue continuous spectrum veiling the normal late-type absorption features may provide an explanation. Perhaps we are concerned in all of these peculiar stars with extensive surface activity in the form of "flares" or hot spots which may resemble the phenomena on the surface of the sun which are responsible for magnetic storms, anomalous outbursts of microwave radiations, etc.

It must be recognized that the variations in light have not been fully explained, even for the more luminous T Tauri variables. However, Herbig's work on RW Aurigae and other similar objects suggests that the process of light-variation may be connected with changes in the character of the dark material that surrounds the star and falls into it.

In RW Aurigae the emission lines change conspicuously in relative intensity with the color-temperature of the star. Moreover, the normal absorption features of late class G are often obscured by a film of continuous emission, as in the central star of B10. Herbig finds that the absorption spectrum is best visible when the brightness of the star is intermediate. At maximum, tenth magnitude, and at minimum, thirteenth magnitude, the absorption features are vanishingly faint.

The presence of this strange continuous emission raises the question whether there may exist some, as yet unknown, process whereby strong ultraviolet stellar radiation (perhaps from the Lyman region) is converted by nebular particles into more or less continuous radiation of longer (and therefore visible) wave lengths. Swings and Struve have called attention to the fact that many solids (as well as liquids and gases) show an intense continuous visible fluorescence when they are excited by far ultraviolet radiation; this fluorescence is sometimes especially pronounced at low temperatures. The specific chemical composition and temperature of the solids, and the specific exciting radiation required for intense fluorescence, may possibly be found in cosmic dust clouds. Such a conversion of ultraviolet energy, in addition to pure scattering, may cause a star embedded in a cloud to appear brighter than would be consistent with the size of the diffuse reflection nebula around it.

An interesting small group of these objects has underlying absorption spectra containing strong Balmer lines of H. These stars are classified as A stars, despite the fact that the character of their emission lines suggest that they are related to the T Tauri variables, and that their association with relatively nearby dark clouds makes it certain that their luminosities are of the order of $+9$. Clearly, these objects are not normal main-sequence A stars, nor are they white dwarfs. But Greenstein has shown that some dark nebulae produce strong absorption lines of H. For example, the spectrum of Hubble's variable nebula NGC 2261, connected with the variable star R Monocerotis, is not of a pure reflection type: at certain times the H absorption lines of the nebula are stronger than those of the star and "there is a suggestion that the long light path through the nebula results in the production of absorption lines." In fact, even the stellar spectrum of R Monocerotis "is difficult to explain . . . as that of a normal stellar reversing layer with superposed emission features" and "the pseudo-supergiant A-type spectrum may be produced in the nebula."

The variations in light of the T Tauri stars are irregular in character. Sometimes they resemble R Coronae Borealis and have long intervals when the brightness remains nearly constant, at maximum, with sudden short drops in light, by one or two magnitudes. Others remain most of the time near minimum and only occasionally brighten up; still others show irregular waves in their light-curves, without preference for either maximum or minimum. Since the changes in brightness are correlated with changes in spectrum, they cannot be caused solely by varying amounts of obscuration in the nebula but must be due to internal processes within the star or within those portions of the nebula which are affected by the star's proximity.

A disturbing thought arises in connection with the ordinary galactic K- and M-type dwarfs, which, though not immersed in nebulous matter, also often show emission lines of Ca II and occasionally those of H. A. H. Joy and R. E. Wilson have catalogued 445 stars having bright lines of Ca II. The distribution of these is shown in Table V.

TABLE V

Distribution of Stars with Emission Lines of Ca II

Spectrum	Supergiants	Giants	Subgiants	Dwarfs
B-A	1	2	0	2
F	4	1	0	4
G0-G4	13	7	0	19
G5-G9	6	15	5	42
K	8	37	9	86
M	5	31	0	124
Total	37	93	14	277

Apparently, the presence of emission lines must be regarded as a symptom which may have several different causes. In spectroscopic binaries it often results from gaseous streams that flow through the system, or from tidal bulges produced by the mutual attraction of the two stellar components. But the dwarfs of Table V are strongly concentrated towards those regions in the sky where there are dense accumulations of cosmic dust. Hence it is possible that the infall of dust particles may account for the bright lines in all of these objects—not only in those which are visibly located in dark nebulae.

In this connection it should be remembered that H. N. Russell has shown long ago that in the case of the sun meteoric particles less than a foot in diameter will be completely volatilized before they reach the photosphere. The vapor will be affected by radiation pressure in a different manner than the solid particles, and an emitting layer of considerable thickness may be formed around the star. The prevalence of the emission lines among stars of lowest luminosity is in favor of the hypothesis. But Russell has also shown that ordinary meteoric matter, of the kind observed to fall on the earth, is insufficient to produce even "the equivalent of a single narrow Fraunhofer line." The nebular material is, of course, much denser than the meteoric matter in the vicinity of the sun. Nevertheless, it must be admitted that the computations require somewhat larger densities than we should like to admit in these clouds. The use of Hoyle's accretion formula, given below, represents a particularly favorable case—that in which the motions of the cloud are very small. The use of an accretion formula less favorable to the mechanism of capture invariably leads to the conclusion that the growth of stars would be a very slow process.

The theoretical work of Hoyle and others makes it probable that the energy radiated by the star in the form of emission lines originally comes from the kinetic energy of the particles of the cloud which are drawn in by the star's gravitation. The process of capturing interstellar particles, by a star passing slowly through a nebula or remaining approximately stationary with regard to its center of gravity, led Hoyle to develop an expression for the amount of matter which falls into the star in each second. If this quantity is multiplied by the potential energy per unit mass which is released during the motion of the matter toward the star, the following formula is obtained for the ratio of the energy radiated by the star in the form of emission lines to its ordinary continuous radiation:

$$\alpha = 10^{-5}N\left(\frac{T}{10^3}\right)^{-3/2}.$$

N is the number of atoms of the cloud per unit volume and T is its temperature. This expression holds for an ordinary dwarf star of spectral type K0, and it ignores the effect of radiation pressure which must be quite negligible. In order to obtain a reasonable value of α, N must be roughly 10^4 when the temperature is also 10^4. This would not be impossible in the case of a dense cloud, so the bombardment of the small and relatively cool dwarf star by interstellar atoms and particles is probably quite ample to produce the observed emission lines.

But it is not probable that the density of the interstellar material is sufficient, at the present time, to permit any large-scale "rejuvenations" of early-type stars to take place. Perhaps there exists some other mechanism which facilitates the infall of interstellar matter into the stars. Magnetic fields could conceivably furnish such a mechanism.

According to the mechanism of internebular bombardment we should expect to observe emission lines only in those cases in which the pressure of the radiation is not sufficient to drive away the particles. An early-type star of the main sequence or a supergiant would have around it a cavity blown free of particles because of the great outward force exerted by its radiation pressure. This explains the observed fact that all ordinary B and A stars within the nebulae have perfectly normal spectra, without a trace of emission. But an early-type star of very small radius would have a large gravitational potential at its surface and might therefore be expected to show emission lines. This may perhaps be the case when the star is of type B or A and is at the same time underluminous to such an extent as to resemble a white dwarf. One of the stars in the Scorpius-Ophiuchus region may be an object of this kind.

There remains little doubt that we are concerned with a real phenomenon of star-growth within a dark nebula, but we do not yet know whether the stars themselves were first produced from the condensation of nebular material or whether they had simply drifted into the nebula from transparent regions of the Milky Way and are now growing in mass, temperature, and luminosity through the bombardment by small particles. Ambarzumian took the view that such aggregations as Joy's group of emission-line stars in Taurus must of necessity have been produced within the nebula itself. This group consists of only a relatively small number of stars and could not remain intact for any great length of time because of the perturbing effect of other stars and of the gravitation of the whole system of the galaxy. He concludes that if these stars are now observed within a relatively small volume of space it must be due to the fact that they were all created at a relatively not very distant time in the past, perhaps not more than 10^8 years ago. But this conclusion is probably not binding. As yet we know little about the motions of these stars with respect to the cloud and with respect to one another. It is not impossible that they originally formed a group of ordinary field stars of the Milky Way not related in any particular manner to each other, and that they simply happened to drift into the nebula, where they are being subjected to the kind of bombardment that would produce emission lines and would at the same time change their physical characteristics.

It is difficult to decide which point of view is more nearly correct. Within these same nebulae we also observe early-type stars near the top of the H-R diagram. It is tempting to believe that all these stars belong somehow together and that the dwarfs will gradually grow in size and mass until they become sufficiently luminous to repel the particles and at the same time radiate as B-type or O-type stars. But this view flatly contradicts the picture accepted by most investigators, according to which a star, when first produced out of the interstellar cloud, resembles an object near the upper end of the H-R diagram, where its mass is large, its temperature is of the order of 20,000° or 30,000° and its radius is 10 or 20 times that of the sun. It is quite true that some of the emission-line stars in dark nebulae are peculiar ob-

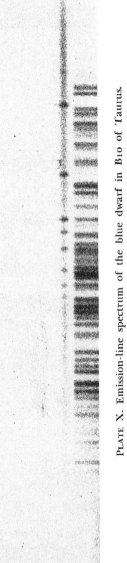

PLATE X. Emission-line spectrum of the blue dwarf in B10 of Taurus.

PLATE XI. J. H. Jeans (1877–1946).

jects, but others seem to be ordinary main-sequence dwarfs with emission lines added to their spectra. Tempting as it is to think of these objects as representing the earliest stage of a star after its condensation out of an interstellar globule, such a conclusion at the present time would be unwarranted. We must admit our ignorance and say that we simply do not know whether these dwarfs were formed within the nebula or drifted into it from the outside.

4. Evolution Through Expansion

Since we cannot be sure whether only early-type giants are formed directly out of interstellar matter, or whether dwarfs may also be formed, we turn to the conclusions of theory. The work of von Weizsaecker, which we shall discuss in detail later on, suggests that a newly formed star is one whose energy per gram and per second exceeds 100 ergs. Such a star belongs to spectral type O or B.

Many stars of these types are even now embedded in clouds of absorbing interstellar matter. Several of them are seen in Plate IX, where they are surrounded by reflection nebulae whose spectra simply duplicate that of the illuminating early-type star. Many years ago E. Hubble showed that stars whose effective temperature is about 20,000°K produce, in addition to these reflection features, strong emission lines of H, He I, [O III], [O II], etc., in the surrounding medium, while the cooler stars almost never produce such nebular radiations. Other groups of O and B stars involved in nebulosity are found in Orion, where they cluster around the great Orion nebula, in Cygnus where they produce a vast network of faintly luminous emission-nebulosities, in Cepheus, in Carina, and in other regions of the Milky Way.

The organic connection between the O and B stars and the diffuse galactic clouds has been firmly established. But not all such stars are now connected with nebulosity. This is not unreasonable, because, as W. Markowitz has pointed out, the group of O and B stars in the Trapezium of Orion would require only about 10^6 years to pass entirely through the Orion Nebula if their velocities with respect to the latter were 10 km/sec.

It is one of the fundamental physical properties of the diffuse clouds in the galaxy that they present visible evidence of a mass of dust or gas in violent turbulent motion. The irregular serrated appearance of the clouds suggests that turbulent motions are far more important in explaining their structure than are the orderly processes of galactic rotation and laminar flow. Since the dark clouds consisting of dust are closely associated with gaseous clouds consisting of H, He, Ca, and other elements, we can frequently measure the turbulent motions of the clouds from the Doppler displacements of the interstellar absorption lines produced in the spectra of very distant stars. Such measurements have been made at several observatories. W. S. Adams has found that the spectra of stars whose distances are several hundred parsecs often contain two, three, or even more, interstellar components of Ca II, Na I, Ti II, etc. These components arise from different turbulent cells within the interstellar clouds and their motions, after correction for galactic rotation, are of the order of 20 km/sec with respect to one another. This is a reasonable estimate of the average velocity of an interstellar turbulent cell.

As these cells condense and interact, rotational velocities must of necessity be set up within the resulting protostars. It is really not surprising that the young stars often have large rotational velocities. It is much more surprising that the interstellar cloud can condense at all without being disrupted by its own rotational motions. However, L. Spitzer, Jr., and F. L. Whipple have shown that internal friction in the original cloud is sufficient to reduce greatly its angular momentum. Even so, it seems reasonable that the newly formed star will possess, on the average, a large amount of rotational momentum.

The problem is to decide how such a rapidly rotating early-type star will evolve if it is left to itself and is not greatly influenced by neighboring stars or condensations of interstellar matter. We must remember that the enormous majority of the stars lie along the main sequence, and that within this band the greatest concentration occurs at the lower end where the temperatures are low, the masses are small and the radii are also small. It looks as though the K-type and M-type dwarfs represent something in the nature of the final stage in the evolution of a large number of stars. This part of the H-R diagram resembles a sink into which many stars drop when their normal courses of evolution have come to an end. We know that these stars generate energy at a very slow rate, so that conversion of H into He requires an interval of time much greater than that during which the galaxy has been in its present condition. At the same time, observations show that these stars are not rotating rapidly. Their rotational momenta are usually quite small, and they do not possess the requisite conditions for rotational evolution. Ambarzumian has emphasized the fact that evolution along the main sequence must be accompanied by a change in mass. Such a change is not possible as the result of the nuclear process. It can only be produced by the loss of mass through ejection as, for example, in a nova, in a P Cygni-type star, or even in an ordinary solar-type star subject to violent prominence activity. It is not now possible to determine the amount of material which the sun loses through prominence activity, because most of the matter raised to high levels above the sun's normal chromosphere falls back into it and is reabsorbed by the sun's atmosphere; but violent explosive prominences are known to leave the sun permanently. Some of them have been observed to possess accelerated motions directed radially away from the sun, and there exists no doubt that solar particles, once they have been lifted high enough above the photosphere, may be accelerated outward by radiation pressure so that they are driven outward and occasionally cause various terrestrial effects, such as the northern lights and the magnetic disturbances associated with solar flares and other disturbed regions on the sun's surface.

It is a little easier to compute the amount of matter which is lost in a P Cygni-type star or in a nova. Ambarzumian and his associates have calculated that stars like P Cygni annually lose approximately 10^{-5} times the mass of the sun. Hence, this type of activity, if it continued without appreciable change, would exhaust the mass of the star in a few million years, at most, but we do not know whether P Cygni-type activity can last for such a period of time. P Cygni itself was a nova which exploded in the year 1600 A.D. and which, after some more or less violent changes in light and spectrum, gradually approached its present condition in which the light is approximately constant and the spectrum shows strong emission lines with violet absorption borders. We associate these absorption borders with outward motion of the gases in the atmosphere. But spectroscopic observations have been conducted for such a short interval that we cannot assume that the process must of necessity last more than a few hundred years. Other stars with similar spectra have shown marked changes, and even the spectrum of P Cygni itself is not always exactly the same.

An interesting example of a star which loses mass by expansion is 17 Leporis. This object possesses an outer layer which expands, on the average, with a velocity of the order of 50 km/sec, but the spectrum undergoes marked variations. In the course of fourteen years, from 1928 to 1942, while this star was under observation at the Yerkes Observatory, nine violent outbursts of a nova-like character were observed in the spectrum. In addition, there were several minor disturbances. These outbursts do not repeat themselves at regular intervals but, on the average, occur once in every 150 days. Each disturbance is characterized by the sudden appearance of violet-displaced absorption components which correspond to a velocity of expansion of about 150 km/sec. For a few days the spectral lines are double. Apparently there are two shells, the original one with a velocity of expansion of the order of 50 km/sec and the new one with a velocity of 150 km/sec. This double-line absorption stage usually lasts only about twenty days. The component lines approach one another and merge to form single absorption lines, as though the violent outburst had gradually been checked by the more moderate expansion of the original shell. Different spectral lines show different intensities in the two expanding shells. For example, the more rapidly expanding shell shows strong lines of Fe II and weak lines of Sc II. Since the ionization potentials of these two ions are 13.6 and 12.8 volts, respectively, we conclude that the ionization in the more violent explosion is greater than in the original shell. Whenever a violet component is formed, the corresponding red component with its velocity of expansion of 50 km/sec becomes weaker and sharper. This is most conspicuous in the case of the lines of Sc II, but it is also present in the other lines. Hence, we must not think of the new outburst as being produced in a new mass of gas hurled into the existing shell as an explosion from a stationary reversing layer. Rather, we must think of the outburst as a sudden acceleration, combined with increased ionization, of a part of the existing expanding shell. The fact that the central intensities of the broad and strong absorption lines of hydrogen and ionized calcium remain near zero during the entire outburst shows that we are not concerned with localized areas, such as the flares on the sun over which violent explosive prominences may be formed, but rather that an entire layer of the expanding shell is suddenly accelerated and at the same time subjected to increased ionization. It is probable that the accelerated layer is at the top, and the physical explanation of the whole process must somehow be related to the building up of great radiation densities within certain spectral lines, perhaps those of the Lyman series of hydrogen, in the deeper layers of the expanding gas.

It would be easy to compute the number of atoms which leave the atmosphere of 17 Leporis per year by measuring the absorptions of the spectral lines and adding up the corresponding numbers of atoms, but such a computation would not be very accurate because there is a possibility that some of the atoms return to the star in a highly ionized state in which they produce no visible absorption lines. Nevertheless, as to order of magnitude, it is reasonable to believe that the shell of 17 Leporis contains a total number of H atoms per square centimeter column which is similar to that contained in the reversing layer of an ordinary A-type star. The result is quite similar to that which was obtained in the case of P Cygni. The loss of mass per year is of the order of 10^{-6} times the mass of the star, so that here, also, the mass would become exhausted in an interval of about 10^6 or 10^7 years. At first sight this process would seem to provide a mechanism by means of which the stars could slide down along the main sequence and gradually become K-type and M-type dwarfs. But it is highly doubtful that P Cygni-type processes of expansion last long enough to convert a massive star into one of small mass. The most compelling argument is derived from the observations. P Cygni stars are not numerous in any spectral class, but they are observed only among those objects which are located near the top of the H-R diagram. Even if expansion of shell-like gases should continue over intervals of time far greater than those which are covered by our observations, the process apparently stops long before a star has been converted from one of type B or type A to one of type K or type M. The dividing line seems to be approximately at type A, where the average masses of the stars are still several times larger than those of the sun. Hence it may be possible for expanding shells to convert a star at the top of the H-R diagram from one lying a little higher to one a little lower, but it is quite impossible to explain in this way the apparent tendency of the stars to accumulate among the dwarfs of later types.

Similar arguments apply also to the ordinary novae. The amount of mass which is blown off in an explosion like that of Nova Cygni 1920 or Nova Herculis 1934 is about 10^{-5} times the mass of the sun. During such an explosion the star increases in luminosity by tens of thousands of times, often in a few hours. The expansion velocity may be as large as 5000 km/sec (in the case of T Coronae Borealis, on February 9, 1946, according to W. W. Morgan and A. J. Deutsch). B. V. Kukarkin and P. P. Parenago have shown that the ordinary novae are probably recurring phenomena. In some stars, like 17 Leporis, the explosions repeat themselves every 150 days or so. In several recurring novae the explosions occur at intervals of some years; in T Coronae Borealis the interval was 80 years. The explosions are not strictly periodic, but the average interval between them is related to the average change in luminosity. If we extrapolate this relation to the ordinary novae, we obtain intervals of the order of a few thousand years. Assuming that at every explosion a normal nova loses $10^{-5} \mathfrak{M}_\odot$ in mass and that the interval between explosions is 10^3 years, we would require 10^8 or 10^9 years to exhaust the mass of the original star. Unfortunately, we do not know what the pre-novae are like. If they are similar to the post-novae, as the work by Kukarkin and Parenago suggests, they are not normal stars. Most of them have continuous spectra, without any lines, but some are characterized by nebular emission lines. There are at least 25 observable novae per year in our galaxy. The total number must be several hundred. There is, however, no reason now to believe that all, or many, stars of the main sequence must of necessity pass through the nova stage in the course of their evolution. The present post-novae are stars of about absolute magnitude $+5$, so that they are not related to the young stars of the H-R diagram.

The Wolf-Rayet stars, which we discussed on page 79, are so rare that they cannot play an important role in stellar evolution. The strange nova-like dwarf variable stars—the SS Cygni and U Geminorum objects—are more abundant, but are probably also exceptions rather than the rule. Binary nature may be responsible for their spectroscopic behavior. We conclude that expansion probably does not greatly change the character of the H-R diagram, with time.

5. Evolution Through Rotational Breakup

The theoretical expectation of von Weizsaecker that young stars should have large rotational velocities suggests another process of evolution which involves a change in mass as the result of rotational breakup. This type of evolution has been studied in great detail by George Darwin and James H. Jeans (Plate XI), as well as by numerous other theoretical astronomers. If a mass of gas is rotating rapidly and its momentum is increased through contraction, the mass may reach a stage in which its gravitation can no longer hold it together. Breakup must then occur, and we do not know what form this process will take in the case of real stars. The classical theory has been discussed by Jeans and applied only to gaseous masses without internal sources of energy. If such a mass is completely concentrated at the center, the envelope containing no mass whatsoever, the resulting configuration is known as the Roche model. If, on the other hand, the substance of the star is incompressible and is distributed with a uniform density throughout the mass, the configuration is known as the incompressible fluid model.

Jeans has investigated the problem of the rotation of a mass of gas or of a fluid, and the manner in which such a mass becomes unstable if the rotation exceeds a certain critical value. He has treated mathematically four different models of a star: (a) the incompressible model, (b) the model of Roche, (c) a generalized Roche model consisting of a homogeneous incompressible core surrounded by an atmosphere of negligible mass and density, and (d) an adiabatic model consisting of a mass of gas in adiabatic equilibrium, in which the pressure, p, and the density, ρ, are connected by the relation $p = k\rho^{\gamma}$, where the quantities k and γ retain the same values throughout the mass.

The incompressible model had already been investigated in detail by Maclaurin, Jacobi, Kelvin, Darwin, and Poincaré. It is now known that when the rotation is small such a mass will be spheroidal in shape, but as soon as the angular velocity exceeds a value ω given by the expression

$$\omega^2/2\pi G\rho = 0.18712,$$

where G is the constant of gravitation, the spheroidal form no longer remains stable and the object, which then has axes in the ratio of 12:12:7, becomes ellipsoidal in form with three different axes. If the rotation is increased still further, the elongation increases until the ratio of the axes is about 23:10:8. At this stage the mass will develop a furrow, giving it the shape of a pear. This structure is not stable, as was first shown by M. Liapounoff, so that a catastrophic process sets in and the body rapidly breaks up into a pair of masses somewhat similar to a double star. Whether this type of fission actually occurs in nature will be discussed in the following chapter. Conceivably, it will be accompanied by the loss of an appreciable fraction of the mass of the original star.

The Roche model has an entirely different character of breakup. If the rotation is gradually increased, the equator of the rapidly rotating mass bulges more and more, until finally a sharp edge is formed and the whole figure becomes lens-shaped. If the rotation is still further increased, the model becomes unstable when

$$\omega^2/2\pi G\rho = 0.36075,$$

and matter begins to be thrown off along the sharp edge of the equator. The reason for this is that the centrifugal force exceeds gravity at the equator. The two other models represent generalizations of these two extreme cases. Jeans has shown that all real bodies must break up by rotation either in a manner which is similar to that of the incompressible body or in a manner similar to that of the Roche model. In the case of the intermediate models, he has investigated under which conditions the breakup will be similar to that of the incompressible body and under which conditions it will resemble the breakup of the Roche model. The details of these mathematical developments are of great interest and significance in astronomy.

He has introduced a parameter s, which designates the ratio of the volume of the atmosphere to that of the core. For the incompressible model $s = 0$ and $\gamma = \infty$, while for the Roche model $s = \infty$ and $\gamma = 1.2$. For the generalized Roche model the dividing line between the two types of breakup occurs when $s = 1/3$, and for the adiabatic model it occurs when $\gamma = 2.2$. For our purpose it is sufficient to realize that a gaseous mass, like a star, presumably can break up either by forming a double star or by shedding matter along the sharp edge of the equator (Figure 17). It is not quite clear which process will predominate, and Jeans' work does not give us a definite answer to this question. He believes that an ordinary star, after having been formed out of interstellar gas or diffuse matter, may be likened to the Roche model. His calculations lead him to suppose that the characteristic function

$$\omega^2/2\pi G\rho = 0.00035.$$

As this primordial star continues to contract, the conservation of angular momentum requires that ω shall increase inversely as the square of the linear dimensions of the star. This results in an increase of the characteristic function which is approximately proportional to the average density raised to the power of $1/3$. Hence, by the time the average density of this primordial star has increased by a factor of 10^9, the value of the characteristic function must have increased by a factor of 10^3. This would bring it to about 0.36, which is the critical value at which a sharp edge at the equator is formed. Any further contraction results in the throwing off of matter at this sharp edge. It is clear that, according to Jeans, this whole process takes place as long as the star is believed to consist of a small dense nucleus and a very large gaseous atmosphere of very low density. We do not

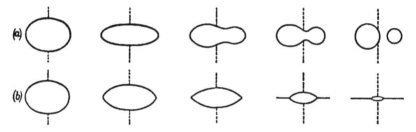

FIGURE 17.   The two evolutionary paths of rotational breakup according to Jeans; (a) masses of nearly uniform density; (b) gaseous masses with high central condensation.

know whether stars are formed in this manner. But let us follow Jeans further. He says that the equatorial ejection of matter will continue until a further critical density is reached at which the star presumably can be regarded as resembling more or less the incompressible model. When that happens the star is ready to become unstable and it will result in forming an ellipsoidal body, ejecting matter only at the two pointed ends.

This ellipsoidal body should, if the rotation is still further increased, give rise to the pear-shaped body which ultimately breaks into two components. According to Jeans, therefore, the final result of the process of disintegration is a binary star. The two components rotate about one another in a more or less dense atmosphere of ejected matter through which they will plow their way. Jeans calls attention to the fact that this formation is reminiscent of J. C. Duncan's hypothesis advanced many years ago toward the explanation of the Cepheid variables. This hypothesis has long ago been abandoned, for the purpose for which it was originally created, but it is of interest to have it revived in connection with studies of close binaries. The transition from the equatorial type of breakup to the fissional type of breakup depends upon the quantity s. A star will first have a very large atmosphere and gradually lose it by equatorial breakup. At the same time, the remainder of the star will condense, forming an increasingly dense nucleus with a relatively small atmosphere around it. Finally, when the ratio of the volume of the atmosphere to the volume of the nucleus is 1/3, the nucleus is ready to break up by the fissional process. We must remember here that throughout this discussion it is supposed that the nucleus remains approximately spherical in shape while equatorial breakup is taking place in the atmosphere. It is therefore entirely possible for this nucleus to reach the critical value of 0.19 for the quantity $\omega^2/2\pi G\rho$ after the atmosphere, for a long time, had been shedding matter by equatorial breakup. In the course of this shedding the star must remain homologous.

6. Stellar Rotation

Let us next see what we can learn from the observations. It is significant that many stars actually have rapid axial rotations. This was not known when Jeans was writing his book, and it is indeed remarkable that so many theories of cosmogony in former years were developed upon the assumption of rotational breakup of one kind or another, without any real evidence that rapid rotations occur among the stars.

The problem of the rotation of stars had its origin, of course, in the study of the sun. The period of rotation of the sun is approximately one month. Its velocity at the equator is approximately 2 km/sec. This is so small a rotation that the sun does not appreciably differ in shape from a sphere. It is exceedingly far removed from the stage of equatorial breakup. However, almost 50 years ago F. Schlesinger found a spectroscopic binary which happens to be an eclipsing variable. The descending branch of the velocity-curve was not as smooth as one would expect from the laws of Kepler. Schlesinger noticed that during the early stages of the eclipse the radial velocities which he had measured on his spectrograms gave systematic departures in such a way as to place the measured points above the smooth run of the velocity-curve. After mid-eclipse the opposite phenomenon took place: the observed points fell below the smooth curve. He concluded that this could be produced if the bright star rotates around its axis so that when it is partly eclipsed by the dark star the remaining portion of its apparent disk is not symmetrical with regard to its axial rotation. It can be easily seen that if the star rotates in the same direction in which it moves within its orbit, then in the early stages of the eclipse the dark body in front will encroach upon that part of the disk of the bright star behind which is approaching the observer. The other, receding, part of the disk remains visible, and consequently the spectral line which before eclipse was symmetrically broadened, both ways, by the rotation of the star, now becomes unsymmetrical: the positive values predominate and the measured point falls above the velocity-curve. If the eclipse is total, then for the duration of the totality we lose sight of the star behind it. But if the eclipse is annular, a ring of the star behind remains visible and in the event of a central eclipse this ring again produces a perfectly

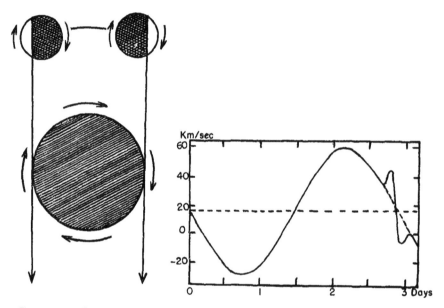

FIGURE 18. Rotational distortion of the velocity-curve of an eclipsing variable during the partial phases of the eclipse.

symmetrical distribution of velocities of approach and recession. Only in this case we have a deficiency of velocities corresponding to zero values. That is, we fail to see a portion of the disk which has little Doppler effect of rotation and, therefore, normally produces only a central deepening of the absorption line. At this stage of the annular eclipse we should observe a shallow absorption line, and in some cases we may even see a line having a central peak, giving it the appearance of a double absorption line. Finally, in the later stages of the eclipse, the obscuring body covers up those portions of the disk of the rotating star which are receding. The approaching portion of the disk is now uncovered. The spectral line again is unsymmetrical and its center of gravity appears displaced toward the violet side of the spectrum (Figure 18). This is a most fruitful and interesting method of studying the rotation of a star. The method was first used with full precision of measurement by R. A. Rossiter and D. B. Mc-Laughlin at the University of Michigan, by the former in the case of β Lyrae, by the latter in the case of Algol.

The same phenomenon has been observed in about 100 systems. It is always present when the conditions lead one to expect that the star has an appreciable rotation and when the eclipse is favorable for producing the necessary asymmetry in the measured lines. In the case of Algol the unsymmetrical shapes of the lines during the partial phases of the eclipse have been recorded with the microphotometer. These shapes are of considerable interest theoretically. It is fascinating to discuss the various shapes of lines which can be produced by assuming different relative diameters of the two stars and different inclinations of their orbits. If the eclipse is central we obtain one kind of asymmetry. If the eclipse is not central we obtain lines of varying shapes. But the study of the individual line contours has not progressed very far because most of the eclipsing variables are too faint to be observed with high-dispersion instruments. For all the rest of the systems we must fall back to the original method of investigation, which depends upon setting the wire of the microscope with which the plates are measured, upon what the observer believes to be the center of gravity of the unsymmetrical line. Fortunately, in doing so the observer is aided by the fact that the spectrograph itself imposes upon the intrinsically unsymmetrical line a contour which is symmetrical. Thus he is not greatly perturbed by the appearance of the line. It looks to him symmetrical, or almost symmetrical, because in most cases the instrumental contour prevails over the true contour of the line as produced by the partly eclipsed rotating star.

In spectroscopic binaries of very short period the lines are always appreciably broadened. In some they are broadened to a very great extent. For example, in Alpha Virginis (Plate XVI) or U Cephei (Figure 19) the lines are several angstroms in width and the velocity of rotation at the equator is estimated to be approximately 200 or 250 km/sec.

The contours of spectral absorption lines in single stars do not differ from those observed in close spectroscopic binaries. There are many single stars, such as Alpha Aquilae or Eta Ursae Majoris, which have dish-shaped lines resembling the lines of Alpha Virginis. It is possible to conclude that in these single stars the line-broadening is also caused by rapid axial rotation.

The rotational velocities which we observe are not the true equatorial velocities but their projections upon the line of sight, $v_0 \sin i$. When $i = 90°$ the axis of the star is at right angles to the line of sight, and we really measure v_0. But when $i = 0°$, the axis lies in the line of sight and we obtain $v_0 \sin i = 0$, and the lines are sharp. In most single stars we cannot determine i. But we shall see that a statistical procedure permits us to derive interesting results from the distribution of the values of $v_0 \sin i$.

The measurements are made by fitting the observed contour of a spectral line to a contour obtained theoretically by applying different

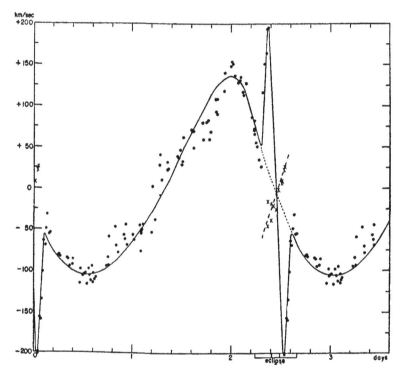

FIGURE 19.   The velocity-curve of U Cephei showing a large rotational distortion during the partial phases of the eclipse. Solid dots represent the velocities of the B8 star. Crosses designate the velocities of the G2 star, which can be photographed only during the total eclipse. The velocity-curve, as a whole, is unsymmetrical, suggesting an eccentricity of the orbit which is incompatible with the photometric investigations.

amounts of Doppler broadening to an intrinsically narrow line-contour having the same equivalent width as the observed line. This is illustrated in Figure 20.

The velocities of rotation at the equator which have thus far been measured in single stars reach, in a few cases, 500 km/sec. It must be remembered that we are not talking here of the broadening of such lines as those of hydrogen in ordinary A-type stars, for example in Vega, but we are discussing a process of broadening which affects uniformly all the lines in a stellar spectrum and in which the broadening is proportional to the wave length, as it should be in the case of a phenomenon caused by Doppler effect.

The kinetic energy of rotation is

$$E = 1/2\omega^2 \int r^2 dm,$$

where ω is the angular velocity of the particle dm, and r is its distance from the center. The quantity $\int r^2 dm$ is the moment of inertia about the axis of rotation. We introduce the quantity k, the radius of gyration, which is defined by

$$k^2 = \int r^2 \frac{dm}{\mathfrak{M}},$$

where $\mathfrak{M}$ is the total mass. Hence, the kinetic energy is

$$E = 1/2k^2\omega^2\mathfrak{M}.$$

For a homogeneous sphere, $k^2 = 0.4R^2$. For a centrally condensed sphere k^2 is smaller. We have, approximately,

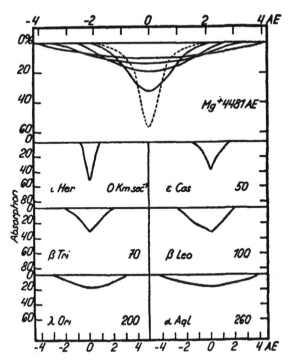

FIGURE 20. Rotational broadening of spectral lines. At the top the dotted line represents the unbroadened line contour, while the other curves represent different amounts of broadening. The observed contours (by C. T. Elvey) are shown below for several stars, with their values of $v_0 \sin i$.

$$E = 0.2\mathfrak{M}R^2\omega^2 = 0.2\mathfrak{M}R^2\left(\frac{2\pi}{P}\right)^2 = 0.2\mathfrak{M}\left(\frac{2\pi R}{P}\right)^2.$$

But $2\pi R/P = v_0$ is the velocity of rotation at the equator. If, for example, $v_0 = 100$ km/sec and $\mathfrak{M} = 5\mathfrak{M}_\odot = 10^{34}$ grams,

$$E = 0.2 \times 10^{48} \text{ ergs.}$$

The central condensation reduces this to about 10^{46} or 10^{47} ergs. This is to be compared with the energy of the rectilinear motion of the star:

$$T = 0.5\mathfrak{M}v^2 = 0.5 \times 10^{34} \times v^2.$$

For a normal B or A star, $v = 10$ km/sec. Hence, $T = 0.5 \times 10^{46}$ ergs. The two amounts are similar in size and are, moreover, comparable to the energy contents of radiation, of translatory motions of atoms and electrons, and of their excitation and ionization. But the available supply of nuclear energy is many times greater.

The most conspicuous correlation between $v_0 \sin i$ and other physical parameters of the stars is with spectral type. The O, B, A, and early F stars frequently have large values of $v_0 \sin i$. On the average the late F-stars rotate much slower, and among the stars of classes G and later rapid rotations occur almost only in close spectroscopic binaries. One star of class G, HD 117555, was found by P. W. Merrill to have ill-defined shallow absorption lines corresponding to a rotational velocity $v_0 \sin i = 75$ km/sec. But this spectrum shows other peculiar features. The $H\alpha$ and Ca II lines are in emission, and the former may be variable. It is noteworthy that large rotational velocities among the late-type stars are so rare; and that in the single instance known at present the H lines are in emission.

Supergiants of early and late spectral classes, and normal giants of classes F and later, never show conspicuous rotations. This is true of Cepheids, long-period variables, and all other groups which fall among the giants and supergiants. Several recent investigators have remarked that in some supergiants and giants the turbulent velocity obtained in the usual manner from the curve of growth is smaller than that derived from the contours of the lines. This effect is present in Epsilon Aurigae according to K. O. Wright and E. Van Dien, in Eta Aquilae according to M. Schwarzschild, and in Delta Canis Majoris according to A. Unsöld and O. Struve. This may be caused, in Delta Canis Majoris, by large-scale turbulence of the order of $v_{\text{turb}} = 22$ km/sec, or by rotation with $v_0 \sin i = 30$ km/sec. Wright and Van Dien also suggested rotation in the case of Epsilon Aurigae. But it is more probable, at present, that we are dealing with turbulent elements whose dimensions are not much smaller than the thickness of the absorbing layer. In any case, the values of $v_0 \sin i$ are small, compared with those observed in the O, B, and A stars.

If the rotational velocities $v_0 \sin i$ are plotted against the galactic longitudes and latitudes of the stars, there is no correlation. We conclude that the distribution of the axes of rotations is a random one. But there are several interesting groups of stars which seem to show either a preferential tendency towards small or towards large values of $v_0 \sin i$. Compact groups with small observed rotations are located in Orion ($l = 175°$, $b = -20°$) and perhaps in Cassiopeia ($l = 90°$, $b = 0°$). Large velocities predominate in the Pleiades, in Perseus ($l = 115°$, $b = -8°$), and in Monoceros ($l = 185°$, $b = -8°$). It is probable that among the stars of such compact groups or clusters there is a tendency for the axes to be parallel. This was suggested by Struve and B. Smith because the rotational velocities of all B-type members of the Pleiades, except one (Maia, whose spectrum is abnormal) are very large. Thus it seems that $i \sim 90°$ for all stars, and there is no reason to suppose that the position angles of the axes are distributed at random, while i is not so distributed. This was confirmed by J. A. Pearce and E. Van Dien. But it is not entirely excluded that all values of v_0 are uniformly very large (about 300 km/sec) but that the inclinations are at random. The F, G and K-type members of the Pleiades have small rotational velocities. There is no correlation between the radial velocities of the stars in our galaxy (corrected for the solar motion) and the values of $v_0 \sin i$.

The lack of correlation between $v_0 \sin i$ and galactic latitude suggests that the axes of rotation are distributed at random. Hence the question arises whether we can assume a uniform rotation for all stars of a given spectral type and attribute the observed variation of $v_0 \sin i$ to the random distribution of the axes. The probability that the inclination lies between i and $i + di$ is proportional to the area of the corresponding spherical zone, or $\sin i \, di$. The probability that the observed velocity is between $v_0 \sin i$ and $v_0 \sin i + d (v_0 \sin i)$ is

$$W(v)dv = c \, \sin i \, di$$

But

$$v = v_0 \sin i$$

$$dv = v_0 \cos i \, di;$$

hence,

$$W(v)dv = c \, \mathrm{tg} \, i (dv/v_0) = c[v/\sqrt{(v_0{}^2 - v^2)}](dv/v_0).$$

The corresponding theoretical distribution has a pronounced maximum at v_0 and is quite different from the observed distribution. We can, however, analyze the observed curve by means of functions of this type and attribute to each a weight that will best represent the observations. The weights will then be proportional to the percentages of stars having the assigned true rotational velocities v_0. The results, taken partly from the work of Miss C. Westgate, are given in Table VI.

TABLE VI

Distribution of True Equatorial Velocities

v_0	0–50 km/sec	50–100 km/sec	100–150 km/sec	150–200 km/sec	200–250 km/sec	250–300 km/sec
Per cent......... B and O	27	53	15	4	1	0
Per cent......... A	16	28	24	20	11	1
Per cent......... F0-F2	30	50	15	4	1	0
Per cent......... F5-F8	70	30	0	0	0	0
Per cent......... G, K, M	100	0	0	0	0	0

The distribution for O, B, and early F are very similar. The A's seem to have a greater percentage of fast rotations, but this result rests only upon Miss Westgate's short series of observations. The startling decline at F5 is unquestionably real. In the later types, spectroscopic binaries like W Ursae Majoris, have been excluded.

There is a remarkable correlation between the rotational velocity as determined in the usual manner from the absorption lines of Mg II, He I, etc., and the total widths of the bright H lines in Be stars which have normal absorption-line B-type spectra with emission lines of H, and sometimes Fe II, superposed over the absorption lines. This correlation is demonstrated in Table VII.

TABLE VII

Correlation between Emission and Absorption Line-widths

Emission Lines	No. of Stars	Estimated Amount of Rotation in Arbitrary Units
Double, average, and single, very broad............	15	8.5
Double, very broad.............................	7	7.3
Double, narrow, and single, average..............	6	5.3
Single, very narrow.............................	6	3.7

Here maximum rotation, arbitrarily designated as 10, corresponds to about 250 km/sec.

Table VIII gives all stars for which R. H. Curtiss had measured the widths of the bright lines:

TABLE VIII

List of Bright Be Stars

Star	Rotation in Arbitrary Units	Width of Hβ
f¹ Cygni...............................	9	10.5Å
π Aquarii..............................	10	7.5
φ Persei...............................	10	7.0
1 Hev. Monocerotis....................	7	6.5
β Monocerotis.........................	7	6.5
b² Cygni..............................	9	6.0
ψ Persei..............................	9	6.0
κ Draconis............................	8	5.0
γ Cassiopeiae.........................	10	5.0
c Persei..............................	5	4.0
β Piscium.............................	1	2.5
11 Camelopardalis.....................	2	1.0

These results leave no doubt that the correlation is conspicuous. It must not be confused with the tendency of B-type supergiants (like Beta Orionis) to show narrow emission lines, often of the P Cygni type.

There has been some discussion whether the widths of the bright lines are caused by rotation. The absence of violet absorption borders strongly militates against expansion as a major effect. But the frequent variations in the relative intensities of the double components of the emission lines suggest, according to D. B. McLaughlin and B. P. Gerasimovič, that expansion and contraction are not negligible. The widths, as measured by R. H. Curtiss, correspond to velocities from 30 km/sec for 11 Camelopardalis to 320 km/sec for f¹ Cygni. These values are, on the whole, not very different from the ones which are found from the absorption lines. But in the best-observed cases, for example that of Phi Persei, the absorption lines give roughly twice the rotational velocity of Hβ. This is consistent with the theory of conservation of angular momentum according to which $v_0 \propto r^{-1}$. The radii of the shells are, according to all methods of study, of the order of twice the radius of the photosphere.

Among the Be stars there are some which show sharp and deep absorption lines of H and some lines of other elements, between the bright components, or even when the latter are not visible. These objects have been designated recently by the term "shells" or "extended atmospheres." The radial velocities of the shells are at least roughly similar to the velocities of the stars as determined from the broad lines of He and from the Stark effect wings of the Balmer lines. Most of the shell spectra are variable in intensity and radial velocity. Some have appeared suddenly, as in Pleione and Gamma Cassiopeiae, others undergo periodic changes, as in Phi Persei and HD 45910, while still others change slowly in the intensities of the "Alpha Cygni" lines, as Zeta Tauri, Epsilon Capricorni, etc.

Relatively few of these stars have been investigated in detail. Almost all have very large rotations, as determined from the absorption lines of He I, and all have very broad emission lines, although not all have strong emission lines.

It appears reasonable to suppose that the gases which give rise to Be-type emission are predominantly present in rapidly rotating stars, and that the correlation between the widths of absorption and emission lines is due to inclination. The frequencies of Table VII suggest that broad emission lines are much more frequent than narrow ones, and the spread in v_0 as observed for ordinary B stars is much reduced. Probably all normal main-sequence Be stars rotate rapidly. If this is so, then the absence of shell spectra among narrow-lined Be stars suggests that the gases are confined to the equatorial plane. It is certainly true that the most conspicuous shells are at the same time the most conspicuous rapidly rotating stars. For Phi Persei, Morgan's small-dispersion spectrograms show $v_0 \sin i \sim 500$ km/sec; and Gamma Cassiopeiae, as well as Pleione, were always known as exceptionally broad-lined stars.

The physical properties of typical shells will be described in the next section. Briefly, they are tenuous masses of gas whose spectrum shows marked effects of dilution and whose radii are of the order of two or three times the radius of the photosphere. The pressures are of the order of 0.1 bar, which is perhaps 10^8 times less than the pressure of their reversing layers. There is usually considerable turbulence, and the rotational velocities are always much smaller than in the reversing layer. In several shells there is a pronounced effect of stratification: thus, in Zeta Tauri the lines of Si II and Mg II originate low in the shell and show appreciable rotation, while Fe II originates higher, where conservation of angular momentum causes the rotational velocity to be smaller. Finally, He I 3965 and H occur at the highest levels, where the rotation is least.

This effect of stratification is also observed in the emission lines of several Be stars. The bright line of Mg II, always weak because of dilution, is usually much broader (when it can be seen at all, as in

Chi Ophiuchi and Gamma Cassiopeiae before the recent outburst) than the bright lines of H and of Fe II. These differences in width can be logically attributed to conservation of angular momentum in a stratified shell. The stratification results from the fact that Mg II is suppressed by dilution at the higher elevations, while temperature-excitation and ionization suppress it in the reversing layer of an early B star, such as Zeta Tauri or Phi Persei. On the other hand, Fe II is favored by dilution. How H will behave is not known. However, it is likely that the pressures in the shells are sufficient to effectively wipe out the metastability of the $2s$ spectroscopic level, so that dilution should produce a weakening of the Balmer lines. The fact that the Balmer lines are strong in spite of this effect in such stars as Phi Persei and Zeta Tauri must be a consequence of the high abundance of H and the fact that turbulence produces an extended flat portion in the curve of growth.

In a few shells there are definite indications of departures from equilibrium conditions produced by the depletion of the continuous radiation of the exciting star on the short-wave side of λ 911. This is true in the case of 14 Comae Berenices, where the absorption lines of H are so affected, and of HD 190073, where the emission lines are weakened.

It is of interest to note that Be stars are unusually frequent in several clusters. In the Pleiades, four out of thirteen stars of types B8 and earlier have bright lines. The ratio is one to three. For the normal B8 stars of the galaxy the ratio is about 1:123, and for B0 to B5 stars it is 1:15. A similar tendency has been observed by Trumpler in the double cluster of Perseus.

7. The Evolution of Pleione

Pleione is one of the brighter stars in the cluster of the Pleiades. Its light is bluish and its spectral characteristics correspond to a surface temperature of the order of 15,000° or 20,000°K. Since we can measure the apparent brightness of Pleione, and since its distance is known to be the same as that of the entire star cluster, we can compute its total output of light. Since, on the other hand, the temperature gives us a measure of the surface brightness, we can divide the former quantity by the latter and thus obtain the surface area and, hence, the diameter of the star. It proves to be about five times larger than the sun.

The spectrum of Pleione has undergone a series of changes since it was first observed in 1888 at the Harvard Observatory. At that time the normal continuous spectrum from the hot photosphere of the star was interrupted by a series of diffuse and very broad absorption lines of hydrogen, helium, and a few other elements. Near the centers of the broad hydrogen absorption lines were narrower double emission components. The other absorption lines had no emission components.

The interpretation of the broad and diffuse absorption lines came when we realized that purely physical causes could not account for the observed contours. The latter were uniformly broad for all elements, without any relation to atomic number, and there was a strong indication that the amount of broadening is proportional to the wave length. This suggested Doppler effect as the correct explanation. It was also noticed that the relative total intensities in absorption (equivalent widths), of members of various multiplets were the same as in other stars whose spectral lines are narrow, the ratios being roughly as follows:

$$A_1 : A_2 : A_3 : \cdots = \sqrt{N_1} : \sqrt{N_2} : \sqrt{N_3} : \cdots$$

If the Doppler effect were of such a character as to influence the absorption coefficient, we should have expected to find for the shallow and broad lines that the relative intensities are proportional to the respective numbers of atoms:

$$A_1 : A_2 : A_3 : \cdots = N_1 : N_2 : N_3 : \cdots$$

Therefore we look for other manifestations of Doppler broadening. It must be remembered that in astronomical practice the point-like image of a star is projected by the mirror or lens of the telescope upon the slit of the spectrograph. We obtain the spectrum of the integrated light of the apparent stellar disk (whose diameter is too small to be resolved). If different parts of the stellar surface have different relative motions in the line of sight, the spectrum will show broadened lines. It is possible to study different types of plausible motions and construct the corresponding line-profiles. Of all such motions, only rotation of the star around an axis produces elliptical profiles resembling those which we observed in Pleione and in many other stars. From the degree of broadening of the absorption lines we can determine the projection, upon the line of sight, of the velocity in km/sec, at the equator of the star ($v_0 \sin i$), where i is the inclination of the axis to the line of sight. In the case of Pleione this quantity is of the order of 300 km/sec. By comparison, we might bear in mind that the corresponding velocity of rotation of the sun is only about 2 km/sec.

It is a remarkable fact that, among the thousands of stars whose spectra have been examined, only a few surpass in equatorial rotational velocity this value of 300 km/sec. It is probable that a dynamical reason exists which prevents an ordinary star from rotating at a greater speed. This reason must be connected with the instability which sets in when the centrifugal acceleration becomes equal to the gravitational acceleration. For our own sun, this point would be reached if the velocity at the equator were made equal to 440 km/sec. But this computation neglects the effect of radiation pressure, which reduces the gravitational acceleration. We can say confidently that the observed upper limit, near 500 km/sec, is in the vicinity of the point where even a massive star would begin to break up through the mechanism of rotational instability.

The existence of emission lines of hydrogen, and occasionally of other elements, in certain stellar spectra is usually attributed to tenuous gaseous masses in the vicinity of these stars. About twenty years ago the hypothesis was advanced that these tenuous masses are ring-like structures, resembling somewhat the rings of Saturn, which revolve around their central stars and therefore produce double components if the observer is located at right angles to the axis of rotation, but only single narrow components if the inclination is 0°.

It is natural that we should associate the existence of tenuous hydrogen rings in such stars as Pleione with the rotational breakup which we inferred as being probably present in stars whose rotational velocity is of the order of 300 km/sec or more.

Statistical studies indicate that the occurrence of hydrogen rings is closely correlated with the measured values of $v_0 \sin i$. The great majority of stars with emission lines of hydrogen have very broad absorption lines. Those few that have narrow absorption lines have also single and narrow emission lines. They are the rapidly rotating stars which we see, approximately, along their axes. Their numbers are quite what we would expect to find statistically if the distribution of the orientations in space were at random.

In any single case where $v_0 \sin i$ is very large and at the same time the emission components of hydrogen are far apart, we have a good chance to find that $i = 90°$. When this condition is realized, the plane of the ring is in the line of sight, and we observe the disk of the star through the gases of the ring. It is possible that absorption lines caused by the ring would then be seen in the spectrum, in addition to the absorption lines from the rapidly rotating atmosphere of the star itself, and the double emission lines would be seen from those two lobes of the rings which are not projected, as observed from the earth, upon the apparent disc of the star.

How should these lines appear? In the first place, we should expect the absorption lines of hydrogen to be devoid of those very broad wings which are produced by ionic Stark effect in the atmospheres of the stars. The rings are presumably so tenuous that the distances between radiating atoms and charged particles would be sufficiently great that the perturbations due to neighboring particles may be neglected. The lines should be exceedingly sharp. Moreover, the rings revolve around the stars with velocities which are two or three times smaller, if measured in km/sec, than the values of $v_0 \sin i$. This, incidentally, suggests that these gases revolve with conservation of angular momentum. We determine these velocities by measuring the separation of the violet edge of the violet component from the red edge of the red component.

Even this rotational broadening would blur the absorption lines, if it were actually effective. In reality, only a small portion of the hydrogen ring is projected upon the apparent disk of the star—that portion which moves almost wholly at right angles to the line of sight. Hence the absorption lines of the ring should show no rotational broadening.

The observational test of this prediction, in the case of Pleione, was delayed by many years. In 1906 the emission lines had disappeared, and from then on until 1938 it resembled an ordinary rapidly rotating star, without indication of rotational breakup. Why the production of emitting rings is an intermittent phenomenon, and not one which takes place continuously, we do not yet know. Dynamical theory predicts that, as the rotational velocity increases, matter will be continually shed at the sharp edge of the equator, and there is no obvious reason why this phenomenon should be interrupted. But it must be remembered that we are concerned with an unstable process, so that small intrinsic changes within the star (similar, perhaps, to the phenomenon of sun-spot production) might well serve to regulate the outflow of gases at the equator.

In 1938 the double hydrogen emission lines reappeared in the spectrum of Pleione. But in addition there appeared a set of very sharp absorption lines, exactly as we have described them previously. The hydrogen absorption cores are seen between the two emission components, and they cut deep into the shallow bottoms of the normal stellar profiles. These lines are seen to about H 30. (In the similar object 48 Librae they have been observed by P. W. Merrill to H 41.)

From the theory of the ionic Stark effect it is possible to predict how many members of the Balmer series will be seen as separate lines, without appreciable confluence of their wings, if the density of the charged particles—in this case that of the free electrons—has a given value. In fact, D. R. Inglis and E. Teller have given the simple expression

$$n_e = 23,26 - 7.5 \log x$$

where n_e is the electron density and x is the number of distinct Balmer lines in the spectrum. (It is assumed, of course, that the resolving power of the spectrograph is sufficient.) In our case this would give

$$n_e = 10^{11} \text{ per cm}^3.$$

However, the theoretical expression was derived under the assumption that perturbations by neighboring particles are the sole cause of the confluence of the upper members of the series. In the tenuous rings which we are discussing, turbulent motions are present which greatly exceed even the thermal velocities. Hence we can only conclude that

$$n_e \leqslant 10^{11} \text{ per cm}^3.$$

Of the utmost interest is the presence, since 1938, of many other sharp absorption lines in the spectrum of Pleione. The chemical composition of the ring must be very similar to that of the star itself (which, in turn, resembles that of the sun), so that the presence of hydrogen alone in emission means only that hydrogen is much more abundant and therefore produces much stronger lines than any of the other elements.

The broad and shallow absorption lines of the rapidly rotating atmosphere of Pleione are still present, and are seen, though not very distinctly, through the gases of the ring. The ring is therefore essentially transparent in ordinary light.

When the absorption lines of Pleione were carefully examined, it was found that in general they resemble with great fidelity the spectra of ordinary supergiant stars that have similar conditions of ionization. There are only a few striking differences: the line of Mg II 4481 and two groups of lines of Si II 4128–31 and 3853–3862 are abnormally weak in the ring of Pleione.

If we examine the energy diagrams of these two ions, and compare them with the energy diagrams of all other elements represented in the ring, we notice a remarkable circumstance. The lines of Mg II and Si II are the only ones whose lower levels are not metastable but possess strong downward transitions which connect them, either in one step or in two steps, with the ground-level. This circumstance brings to our minds the theory of cycles which S. Rosseland proposed twenty years ago, and which had already scored a triumph in connection with the problem of the forbidden lines of the nebulae. It would take too long to explain this theory here in detail. But an analogy may help us appreciate its enormous power in dealing with problems of astronomical spectroscopy.

The atoms of a gas act essentially as a light-filter which reduces the intensity of the transmitted beam of light. A piece of smoked glass, for example, acts in a similar manner. Now, an ordinary absorbing screen, like the smoked glass, always reduces the amount of light by a definite fraction, irrespective of the intensity of the light observed. But suppose that we could make an absorbing screen which had the property (through chemical action) of absorbing very little light if the source is far away, and therefore quite dim, but which could absorb much light if the same source were quite close, and therefore very brilliant. Such a screen would tell us at once, from the amount of light it absorbed, whether the source was close to the screen or far behind it. The atoms of Mg II and Si II act precisely in the manner of this hypothetical screen. They absorb much light, and produce strong absorption lines, if they are immediately above the radiating surface of the star, in its own atmosphere; but they absorb little light and produce weak absorption lines if they are located in a ring far above the surface of the star. All the other atoms of Pleione (with the exception, perhaps, of hydrogen) lack this remarkable property. They tell us nothing about the relative location of star and absorbing gas. From the intensities of the lines of Mg II and Si II we can even determine how far above the surface the ring is located. The result is two or three times the radius of Pleione.

Since 1938 the spectrum of the ring of Pleione has continued to evolve. At first there was a gradual strengthening of all absorption lines, without any striking change in the relative intensities of lines of different elements. The amount of matter in the ring (per cm^2) increased by a factor of about 1000 in the interval from 1938 to about 1941; since there was no corresponding decrease in ionization, the ring must have changed in thickness rather than in density.

From 1940 to 1949 there has been a slow but steady decrease in ionization, followed by a rapid increase in ionization in the last year, when all lines began to fade. We tabulate here, for each year, the characteristic ion whose lines were especially strong, together with the corresponding ionization potentials:

1938–1940	Ni II	18.2 volts
1941	Cr II	16.6
1941	Fe II	16.5
1942	Ti II	13.6
1943	Mn II	15.7
1944	Sc II	12.8
1945	Sr II	11.0
1946–1948	Fe I	7.8
1949–1950	Fe II, Cr II	16.5

The lines of hydrogen have been omitted in this tabulation. They were at all times much stronger than the lines of the other elements. Until 1943 or 1944 they had become gradually stronger, but since 1945 or 1946 they have started to get fainter.

The tabulation of the elements shows a gradual change towards lower ionization, in excellent accord with the behavior of hydrogen, but much too slow to account for the thousand-fold change in the total numbers of atoms per cm². There was only one very remarkable anomaly: the relatively late strengthening of the lines of Mn II. The ionization potential, as well as the relevant excitation potentials, of the lines of Mn II are intermediate between those of Fe II and Ti II. Yet there can be no doubt that the lines of Ti II became strong at a much earlier stage in the evolution of the ring than the lines of Mn II.

This discrepancy is only one of several similar which have been noticed in Pleione and in a number of other objects. It may seem at first glance to be a rather unimportant detail that might be cleared up after the excitation and ionization of the gases in the ring are better understood, and it is possible that it may be caused by the statistical weights. But it is probable that this simple explanation will not be sufficient.

Another interesting question is that of the character of the ionizing radiation. We know that we can see the real star through the gas of the ring. Hence it would seem reasonable to suppose that it is the radiation of this star which produces the ionization and excitation of the atoms in the ring. But what do we know about this radiation? The effective temperature of Pleione is about 20,000°K. But is its energy-curve similar to that of a black body in those wave lengths which contribute to the phenomena of ionization and excitation? The ionization energies, for example, which are listed in the table lie on both sides of λ 911, the limit of the Lyman series of hydrogen.

There is very little prior information to guide us. Direct observational data on stellar energy distributions reach to about λ 3000, where the ozone and oxygen absorptions of the air abruptly cut off the spectrum. Indirect results have been obtained from studies of gaseous nebulae by assuming the stars to radiate as black bodies, and fairly reasonable values of their temperatures have been derived in this manner. On the other hand, in the case of the sun we have a varied assortment of facts, all of which seem to demand an enormous excess of extreme ultraviolet radiation above that which would be present in a black body of the sun's effective temperature. We cannot be certain, therefore, that a temperature of 20,000°K in the case of Pleione, with a black-body distribution of energy, will give correct results when introduced into the ionization equation of Saha. More difficult still is the question of possible departures from the black-body curve of the stellar-energy distribution. Because of the enormous preponderance of hydrogen in the atmospheres of all stars, the continuous absorption at the Lyman, Balmer, Paschen, etc., limits of hydrogen are especially effective in producing the general opacity of a stellar atmosphere like that of Pleione. The result is an energy distribution which bears only a distant resemblance to Planck's curve. It has abrupt "jumps" at the various hydrogen limits. Especially at the Lyman limit, the energy on the violet side of λ 911 should be many times smaller than on the red side. The abruptness of this "jump" is somewhat exaggerated in recent theoretical calculations, because all of them (the most recent ones are by the Russian astrophysicist E. R. Mustel) neglect the effect of absorption in the Lyman lines. These will tend to have very broad and strong wings, merging together near the limit, so that the energy-curve on the red side of λ 911 will appear less abrupt than the published curves have shown. Nevertheless, we should expect in the region between Lyman alpha and beta a considerable excess of radiation over that corresponding to a black body of 20,000°.

The observations show little effect of any possible discontinuity at the Lyman limit. If it were present, we should expect to observe a suppression of second-stage ionization of those elements whose ionization potentials lie above that of hydrogen, 13.5 volts, and a corresponding strengthening of the lines of Ni II, Cr II, Fe II, Ti II, Mn II. Those elements whose potentials are less than 13.5 volts would be doubly ionized to a greater extent, and this would give rise to a suppression of such lines as those of Sc II, Sr II, Ca II, etc. It is possible that a slight effect of this sort is present: for example, the lines of Ca II are relatively faint in Pleione. But in view of the gradual change of ionization, it is difficult to be certain. We can say definitely that the effect cannot be large. As an estimate we might suggest that it cannot exceed a factor of ten in the departures of the energies from the black-body curve. This limit is too crude to permit any further deductions.

Somewhat more revealing is the question of the average temperature of the ionizing radiation for the group of elements in our table. Consider Ni II, for example. Its lines were exceptionally strong at one stage, and they remained conspicuous even when the ionization had been somewhat reduced. The ionization potential is 18.2 volts. Saha's equation of ionization equilibrium for conditions of thermodynamic equilibrium is

$$\log \frac{n^{++}}{n^{+}} + \log n_e = -I \frac{5040}{T} + 1.5 \log T + 15.4,$$

where n^{++} and n^{+} stand for the numbers per cm³ of doubly ionized Ni and of singly ionized Ni atoms, I is the ionization potential, and T is the temperature. If the gas is not in thermodynamic equilibrium, but is being illuminated by a source whose energy distribution is that of a black body, but which subtends an angle ω as seen from a repre-

sentative point in the gas, then the number of photoelectric ionizations is reduced by a factor of $W = \omega/4\pi$, while the number of re-
combinations is not altered. Hence, the equilibrium condition is
now:

$$\log \frac{n^{++}}{n^+} + \log \frac{n_e}{W} = -I\frac{5040}{T} + 1.5 \log T + 15.4.$$

In order to observe strong lines of Ni II, the ratio n^{++}/n^+ must be
of the order of 0.1 to 1.0, because otherwise the ionization would
either be too great, so that most Ni atoms would be doubly ionized,
or it would be so low that most Ni atoms would be neutral. Since we
are concerned only with rough order-of-magnitude estimates, let us
assume that $n^{++}/n^+ = 1$. Then, inserting $I = 18.2$ volts, and $T = 20,000°$, we find

$$\log \frac{n_e}{W} = +17.2.$$

The quantity $W = \omega/4\pi$ is closely related to the results we have derived by means of the theory of cycles as applied to the lines of Mg II
and Si II. It is found that W cannot be less than 0.01. (It will be noticed that the maximum value possible is $W = 1$, which corresponds
to exact thermodynamic equilibrium.) Hence, we find

$$\log n_e \geqslant 15 \text{ or } n_e \geqslant 10^{15}.$$

But earlier, in discussing the sharpness of the hydrogen absorption
lines, we had concluded that

$$n_e \leqslant 10^{11}.$$

The two estimates are irreconcilable. We accept the smaller as more
reliable. Moreover, an electron density of 10^{15} would be much
greater than that of a normal stellar atmosphere, which is impossible,
in view of the tenuous character of the rings. Hence we must question the values used in the ionization equation, and the most doubtful of these is T.

If we had arbitrarily used $T = 10,000°$ in the ionization equation,
then our result for n_e would have been 10^{10}—in good agreement with
the limit obtained from the hydrogen lines.

The question is now: why is T so low? It is not a matter of the
shape of the energy-curve, because other elements, like Sc II, would
have given us substantially the same result. The answer is probably
that we must revise our ideas concerning the physical character of the
ring. We had assumed it to be transparent because we could see the
lines of H and He of the rapidly rotating star through it and had
therefore used the temperature of the star itself ($T = 20,000°$) in the

ionization equation. But if we examine the problem closely, we see that the ring is probably partly opaque beyond the Balmer limit at $\lambda\,3647$, and that it must be completely opaque, not only in the region $\lambda < 911$ Å, but even on the red side of the Lyman limit where the strong and broad Lyman lines crowd together. The radiation of the star reaches only the innermost layers of the rings and is then quickly converted into continuous radiation of the ring itself, just as the outermost layers of a star convert the radiation from the hot interior. To compute the ionization of elements like Ni II or Sc II we must therefore use the temperature that is characteristic of the ring itself. Furthermore, we must use a value of W close to 1, because now the source of the ionizing radiation is in the ring and $\omega = 4\pi$. We see readily that under the assumption that $n^{++}/n^+ = 1$, we now require about $T = 8{,}000°$ in order to obtain $n_e = 10^{10}$.

It is possible to give a number of reasons why n_e should not be much less than 10^{10} (for example, the metastability of the $2s$ level of hydrogen), but we shall not go into these details.

We must, however, clarify the meaning of $W \sim 0.01$, which we found from the Mg II and Si II lines. The energy required to excite the lower (non-metastable) levels of these lines is about 9 volts, so that the radiation comes from the region of $\lambda\,1400$. At this wave length, then, which is well to the red of Lyman alpha, the ring is not completely opaque and the radiation of the star shines at least partly through it. It is quite possible that an even smaller value of W would have been obtained for a line whose energy of excitation comes from a wave length on the red side of the Balmer limit at $\lambda\,3647$. Unfortunately, there is at present no such line available to test this question.

The ring of Pleione is nearly transparent, outside the lines, in the ordinary photographic region. If we had at our disposal a super-telescope with which we could resolve the angular diameter of the ring, we should be able to observe the undimmed light of the star through it. But with an ultraviolet filter cutting off all wave lengths greater than 3647 Å, we should see only a faint image of the star within the luminous outlines of the ring; finally, with a filter transparent to wave lengths less than 911 Å we should not see the star at all.

Anne B. Underhill has called attention to the unsymmetrical contours of the sharp absorption lines of Pleione. She has explained these contours as the result of differential motions within the ring-like shell. The inner layer of this shell rotates with $v_0 = 140$ km, while for the photosphere of Pleione $v_0 = 300$ km/sec. The outer layers of the shell rotate slowly, but they contract and expand with a velocity of about 12 km/sec, in a period of about four months.

There is as yet no way to find out how thick the ring is, at right angles to its principal plane. Since its absorption lines are very strong, we must conclude that the stellar disk is completely covered by the projection of the ring. On the other hand, there are indications that the rings have radii several times larger than their thicknesses. We know little about the distribution of matter along a radius in the equatorial plane of the ring. Indirect results suggest that the inner edge of the ring is usually at a considerable distance from the atmosphere of the star itself. The spectra of the rings of Pleione and 48 Librae are shown in Plates XII and XIII.

8. The Cosmogonical Theory of C. F. von Weizsaecker

It is probable that the ring-like structures which we have observed in Pleione, in 48 Librae, etc., actually represent the process of rotational breakup which Jeans predicted for centrally condensed stars similar to the Roche model. But there remain two difficulties:

(1) The observed rotational velocities of stars which produce rings, though large, are not always sufficient to balance gravitation by centrifugal acceleration. Since the gravitational acceleration is

$$g_1 = 2.74 \times 10^4 (\mathfrak{M}/R^2),$$

and the centrifugal acceleration is

$$g_2 = 1.44 \times 10^{-11} (v_0^2/R),$$

we find that, when $g_1 = g_2$,

$$v_0 = 4.4 \times 10^2 \sqrt{\mathfrak{M}/R} \text{ km/sec.}$$

Thus, for a B star with $\mathfrak{M} = 20\odot$ and $R = 5R_\odot$, we have $v_0 = 900$ km/sec. This is about twice the maximum value observed for $v_0 \sin i$. Perhaps radiation pressure cuts down the value of g_1 sufficiently to account for the difference.

(2) The theory of Jeans assumes that the stars are contracting, thereby increasing the value of the critical quantity $\omega^2/2\pi G\rho$. But we now believe that stars in the main sequence are fed by the carbon cycle of nuclear transformations; if this process were operating alone the stars would not contract, but expand. Contraction is probably important in the protostars and globules of interstellar dust, but it plays no role after the stars have attained a central temperature of 2×10^7 degrees.

Loss of mass by ejection, as in a nova, will result in a gradual shrinkage of the star, but the shell-spectra are often associated with stars which have no P Cygni-type lines.

A solution of this difficulty may come from the fact that large turbulent velocities are always present in the rings of rapidly rotating stars. It is precisely this feature which forms the fundamental property of a new cosmogonical theory which was published by von Weizsaecker in 1947.

In its essential features this theory is a sequel to his earlier paper on the origin of the planetary system. He has given special consideration to the problems of turbulence in cosmic masses and to the effect of rotation in its relation to the evolution of the cosmic bodies. The purpose of the theory is to describe the origin of star systems and of individual stars out of diffuse masses of gas. He makes the fundamental assumption that all star systems and stars have originated within an interval of time which may be described as the age of the world and which can be estimated to be between two and six times 10^9 years. During this interval of time it is believed that the present laws of physics have been effective. No new laws are postulated. During this entire interval of time the total mass of the accessible universe has remained constant, except for a relatively small transformation of mass into radiant energy. Before the stars were formed, the cosmic system is believed to have consisted of a diffuse gas, the chemical composition of which must have been essentially similar to that which we observe today. The formation of the complex atoms, especially those whose atomic weights are greater than that of oxygen, is supposed to have taken place before the beginning of the present state of the universe. It is believed that the primordial mass of gas possessed large motions of a turbulent character in its various parts. Small concentrations within the original gas ultimately condense into stars which from the beginning possess large amounts of rotational momentum. The character of the motions within the newly formed star must be quite complicated. Certainly the star is not rotating as a solid body. In its outermost portions, the rotational velocities are given by Kepler's third law. In the inner regions, friction is so large that the core would rotate approximately as a rigid body. In the intermediate layers, turbulent motions are developed which carry angular momentum with them. A rotating gaseous mass surrounding a nucleus which ultimately may become a star like the sun would break up into turbulent cells whose dimensions are roughly one order of magnitude smaller than the diameter of the original nebulous mass. These turbulent cells are moving about in different directions. Some of them sink toward the central regions, with a loss of angular momentum and with a transfer of energy to other cells which would enable the latter to escape and carry with them the acquired angular momentum. Such turbulent cells may actually escape into infinity. The mechanism of this process consists in the following: the turbulent inner friction of the deeper layers of the gaseous mass causes the latter to rotate more slowly, while the outer regions become accelerated so that there is a continuous transport of energy and angular momentum from the inside to the outside, and the outside portions gradually escape to infinity.

In this manner the radius of the formation becomes smaller and smaller. Von Weizsaecker has shown that if the radius of the mass becomes one-half of what it was originally, then it would be sufficient to let approximately one-half of the mass occupy the smaller inside volume while the outer half acquires a sufficient amount of energy to escape to infinity. However, if the radius of the formation should become reduced by a factor of 100, then it is sufficient that one-tenth of the mass should finally become concentrated within the inner portion, leaving the remaining nine-tenths of the mass free to escape to infinity. In this manner the inner part of the mass loses most of its rotational momentum. It is not necessary to conclude that the angular velocity also decreases, because it is known that the angular momentum with constant angular velocity decreases as the square of the radius. Von Weizsaecker assumes that a star having become sufficiently concentrated to set into operation the nuclear processes of energy generation would still retain a large angular velocity of rotation. Therefore it would not be secularly stable and would attempt still further to get rid of excess angular momentum. Gradually this process of shedding outside regions which carry off part of the rotational momentum will become so effective as to leave the star in a stable state. It may, for example, be supposed that the sun has reached this state.

Von Weizsaecker has given a simple expression to estimate the length of time in years during which a star may develop in the manner described. Let us assume that the original radius of the turbulent mass is r_{max} and the final radius of the star after having shed most of its angular momentum is r_{min}. If l is the mixing length of the largest turbulent elements, and if v is the average radial velocity of these turbulent elements, then the time that it takes for this process to be completed is

$$T = \frac{r^3_{max}}{r_{min}lv}.$$

It is difficult to estimate the quantities which enter into this expression. Moreover, von Weizsaecker states that the derivation of the formula is possible only under greatly simplified assumptions. However, he points out that it rests almost entirely upon dimensional considerations. The data which describe the initial condition of the system are r_{max} and v, because the mixing length l cannot be considered to be independent of r_{max} but must be approximately one order of magnitude smaller than r_{max}. The only quantity having the dimension of time which may be constructed out of these is r_{max}/v. This is the length of time in which a turbulent element moving with a uniform velocity, v, would require to pass from the center of the system to the boundary. But the statistical nature of the process, which involves the origin and disappearance of the turbulent elements, causes a slightly greater length for the transport of the energy, and this is indicated by the factor r_{max}/l. The product of these two ratios would give the length of time required to transport, by means of turbulence, a quantity of energy comparable to the original energy content of the system from the inner portions of the system to the boundary. However, if r_{max} is large compared to r_{min}, then the total energy which is gained by the shrinkage and which must be transported to the outside, will be measured by the ratio r_{max}/r_{min}, and this increases the length of time by the same factor.

More recently, von Weizsaecker has published in the *Zeitschrift für Naturforschung* (1948) the mathematical treatment of the turbulent processes involved. It is of interest to apply this formula to obtain a rough estimate of the time T for a typical star. Let us take a star of early type, O or B, on the main sequence. We assume that the original size of the nebulous mass, r_{max}, was approximately equal to the radius of the orbit of Neptune, 5×10^{14} cm. The quantity r_{min} might be set equal to 5×10^{11} cm, a quantity which results from the temperature and the luminosity of a typical O-type star. For the quantity v, von Weizsaecker sets, rather arbitrarily, 1 km/sec, a value he had derived in connection with his theory of the origin of the solar system. Observationally, we know that turbulent motions are present in the rings and that sometimes $v_{turb} = 20$ km/sec, or even more. But 1 km/sec may serve as an estimate of the average velocity. We shall apply it to the O-type star. Assuming, next, that the ratio $r_{max}/l = 6$, which is the quantity derived theoretically for the nebulous mass around the sun, we find $T = 10^6$ years. This is much shorter than the age of the universe. It is, however, similar to the age of an O-type star obtained from the Bethe cycle of energy generation, or 10^7 years.

A normal O-type star must not be supposed to have completed its process of freeing itself of the excess rotational momentum. The fact that we observe some of these stars with large rotational velocities of the order of more than 250 km/sec shows that the end has not yet come, and we probably observe this process at work in the rings of Pleione and other similar shell-stars. The theory does not make a distinction between different types of shedding matter with consequent loss of rotational momentum. It would be possible for a star to lose matter along the equator, forming rings around rapidly rotating stars. But it would also be possible for a part of the rotational momentum to be used up in the formation of a binary system. The

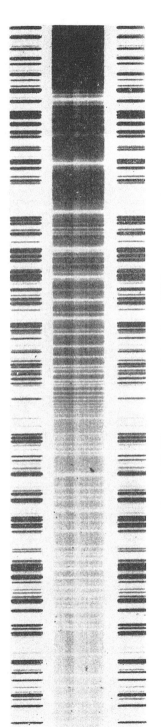

PLATE XII. The spectrum of the ring of Pleione.

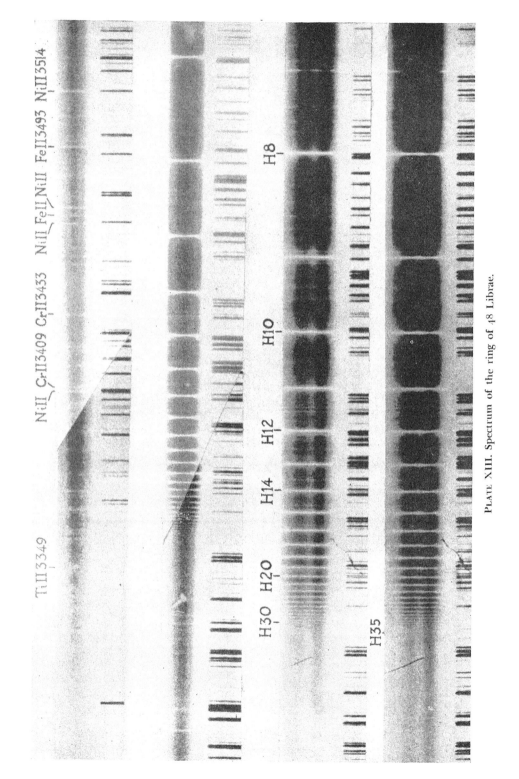

PLATE XIII. Spectrum of the ring of 48 Librae.

existence of close spectroscopic binaries may suggest that such a process is really at work. In any case, we regard it as certain that the O-type and B-type stars are relatively young objects in our galaxy. There are four arguments in favor of this conclusion. First, they are generally considered to be rapidly rotating objects, and it is the principal result of von Weizsaecker's work that rotation forms a relatively early stage in the evolution of a cosmic object. Second, it is probable that those types of stars which occur predominantly within spiral arms of galactic systems are relatively young in the scale of time defined in this work. They are the ones which, according to Baade, constitute Population I. They are also the stars which produce the usual H-R diagram. Stars which belong to Baade's Population II, and give the H-R diagram of a globular cluster, must be regarded as relatively old. The justification of this assignment rests partly upon the work of Baade, who had found that in the local swimming hole of our galaxy these stars are high-velocity objects, having come into the neighborhood of the sun from the central regions of the galactic system. They are relatively old, because they come from regions where there is no diffuse matter left in the form of bright and dark nebulae. Third, we notice that the O and B stars show a strong concentration toward the galactic plane, and that they are intimately connected with the layer of gaseous matter in the spiral arms of our galaxy because this layer is also strongly concentrated toward the central plane. And fourth, they radiate energy at a rate of more than 100 ergs per gram and per second, and will exhaust their supply of H in a relatively short interval of time.

Although the mechanism of turbulence will, in time, carry off a fraction of the star's original rotational momentum, Ter Haar has shown that if the central star has a rotational velocity at the eauator $v = 100$ km/sec the drag of the gaseous envelope can produce only a small amount of dissipation of the angular momentum. It is apparently necessary that there should be in operation some process which continuously transfers momentum from the star to the disk-shaped envelope. In reality, this mechanism may reside in the prominences which, at least in some close binaries, tend to produce just such a transfer. But Ter Haar has suggested a more general process, namely the interaction of the magnetic field of the star and the regions of interstellar ionized hydrogen. Originally suggested by Alfvén in connection with the solar rotation this process results in transferring the rotation of the star to the interstellar cloud. Consisting of many charged particles, such a cloud can be thought of as rotating in a magnetic field within a frame of reference which rotates with the star. This sets up electric currents in the cloud which slow down the rotation of the cloud in this frame. In the original frame of reference the star rotates less rapidly and the cloud is set into rotation. This process may explain why the younger stars have not yet lost much of their original angular momentum, while the cooler dwarfs have been slowed down almost completely.

9. Two Evolutionary Paths

The theory of turbulence accounts, in a remarkable manner, for the observed fact that the early-type stars usually have rapid rotations, while those of types G and later have none. But statistical studies of stellar rotations show that there are a considerable number of stars of types O and B which rotate slowly. As yet, we cannot be certain that every individual star whose absorption lines are sharp has a slow rotation; but it seems probable that such objects as 10 Lacertae, Tau Scorpii, and Gamma Pegasi are really devoid of much angular momentum and are, therefore, in no danger of losing mass either by the shedding of atoms at the equator or by splitting into double stars. This is only a tentative suggestion, but it must be remembered that ordinary O-type stars are rarely free of emission lines. The usual appearance of an O-type star is that of a continuous spectrum with extremely few, shallow absorption lines, often broadened by rotation, but sometimes fairly narrow. Ordinarily these lines are suggestive of large turbulent motions. The line He II 4686 and the pair of N III lines at 4634 and 4641 are often in emission, as is also Hβ and, occasionally, the lines of Si IV. The more conspicuous stars of this kind are usually designated as Of objects. Some of them, for example 9 Sagittae, have fairly narrow lines, although rotation is by no means negligible in these objects. Apparently it is true that we can observe that there are emission lines, even when the axis of rotation is not too far removed from the line of sight. On the other hand, 10 Lacertae contains no noticeable emission of any kind, nor do the lines indicate turbulent motions. It is almost certain that the spectra of slowly rotating O and B stars differ systematically from the spectra of rapidly rotating stars, even though the latter may be oriented in such a way that the component in the line of sight of the rotational velocity is zero. It must be pointed out that stars like 10 Lacertae and Tau Scorpii are really quite rare. No estimate of their frequency in space has been obtained, but they are usually somewhat less luminous than the average star of the same spectral type. Until we have information to the contrary, we shall assume that they have had the same past history as have the rapidly rotating O and B stars. But in the future their life histories must diverge. The rapidly rotating stars are losing mass by turbulence, by equatorial breakup, or by some such process as P Cygni-type emission, while the slowly rotating stars are left to themselves and remain in their present location within the H-R diagram until the nuclear processes have changed the chemical composition of their interiors sufficiently to alter their luminosities and internal structures.

It is not definitely known how a star will move in the H-R diagram if its evolution is entirely that of the conversion of H into He. The mass remains approximately constant. Consequently the star would be expected to evolve along a line of constant mass, but this line is not necessarily identical with the line which we have drawn in Figure 8 from the mass-luminosity relation, because, as we have seen, the absolute magnitude of the star increases as its hydrogen is used up. At the same time, the radius will also increase and the effective temperature will decrease. The star must move away from the main sequence, but exactly where it will be located at every intermediate stage of hydrogen content is not known, because the structure of the star also undergoes a change which is as yet not entirely clear. Consequently, we can only say that a star which produces energy in accordance with the Bethe-Weizsaecker process will increase its radius by an amount which is much too small to accord with the observations. In other words, the predicted change in radius is too small to explain the observed radii of the K-type giants and subgiants. Hence it is necessary to adopt a different model for such a giant than the one which had been adopted by Eddington and others for the stars of the main sequence. A. Reiz and, more recently, R. S. Richardson and M. Schwarzschild have suggested different types of models with convective cores and envelopes in radiative transfer, which will explain the observations and still retain the central temperature of approximately 20 million degrees which is required for the Bethe-Weizsaecker process.

We have already found that it may be best to adopt the results of the observations from the H-R diagrams of clusters. The various groups differ from each other in the extent of the main sequence at its upper end. Within each cluster the stars near the left-hand upper end are anomalous in luminosity and in spectrum. Irrespective of whether they are B-type stars, as in the Pleiades, or A-type stars, as in the Hyades, it seems as though the evolutionary trend among the B-type stars is to slide along the main sequence, leaving only a few objects which are transformed into red giants by the Bethe-Weizsaecker process and which are perhaps ultimately reduced to the condition of a white dwarf, by the process of contraction. Thus, the path of a star like 10 Lacertae presumably would be along a slightly inclined line toward the right side of the diagram, giving it a greater luminosity and a lower surface temperature, as well as a greater radius, than when it occupied its original position in the H-R diagram. Other workers have suggested that if the process continues, after all the H has been used up, such a star would rapidly cross the H-R diagram diagonally from the upper right to the lower left and would end up as a white dwarf. We know that the white dwarfs have masses of the order of one solar mass. Hence, if this evolution is possible, there must be a loss of mass between the red giant and white dwarf stages. Moreover, the white dwarfs are quite frequent in space. They, too, can be regarded as forming a sink which collects the end products of the evolution of slowly rotating stars. The process of conversion of hydrogen into helium requires approximately 10^7 years for an O-type or early B-type star. This interval of time is similar to that required to transform by rotational breakup a star of spectral class B into a star of spectral class G, K, or M. Perhaps this is significant, since the two processes will compete with each other. When the rotational momentum is small, the evolution will be of the type we have just been discussing; when the rotation is large, the evolution will be along the main sequence and the star will end up among the red dwarfs.

The white dwarfs form the second most numerous group of stars in the normal H-R diagram. Roughly speaking, the ratio per unit volume is 100/1 in favor of the main sequence. This is interesting, in view of the fact that only approximately 10 per cent of the early-type stars show slow axial rotations. If, in addition, the period of time required for a star to pass from type B through giant K into the white dwarf sequence is ten times longer than that required for a star to pass from one end of the main sequence to the other, that might explain the observed preponderance of late-type dwarfs over the white dwarfs.

10. Conclusions

There is one fundamental weakness in the von Weizsaecker theory as applied to the evolution of single stars. The nebulous envelopes in which the turbulent processes take place must be observable. We have seen that among the early-type single stars they are actually observed in the form of "shell-spectra." But these spectra are never seen among the stars of spectral classes later than A, and they are almost always associated with stars whose normal spectra are O and B. The latest spectral type associated with a shell is A5, in the case of 14 Comae Berenices. The spectrum of its shell is about F. It would, therefore, seem that the turbulent mechanism ceases to operate in the early subdivisions of class A, where the average rotational velocities are still large. Yet the important thing is to explain the sudden drop in average and maximum rotational velocity at about F5. Table IX illustrates this. The data are the best estimates now available. The first four values for the average velocity are from a statistical discussion by Chandrasekhar and Münch. They suggest that there may be a process which reduces the large velocities of about 500 km/sec to maximum values of the order of 250 km/sec in the absorption stars of classes O, B, A, and early F. This may be the von Weizsaecker process, as observed in the form of rings. But there must be still another process which accounts for the sudden drop at F5.

TABLE IX

Equatorial Velocities of Rotation

Spectrum	Mass	Average Equ. Rot. Velocity	Maximum Equ. Rot. Velocity
Oe-Be........................	15$\mathfrak{M}_\odot$	350 km/sec	500 km/sec
O-B.........................	15	94	250
A...........................	4	112	250
F0-F2.......................	2.5	51	250
F5-F8.......................	1.5	20	50
dG..........................	1.2	0	0
dK..........................	1.0	0	0
dM..........................	0.5	0	0

In the preceding discussion we have attempted to identify the rings of Pleione, Gamma Cassiopeiae, Zeta Tauri, and other similar stars with von Weizsaecker's hypothetical nebula in which turbulence is set up by differential motions. But there is a great difference in the density of this nebula, for which von Weizsaecker adopts a mass of 0.1$\mathfrak{M}_\odot$, and our observed rings, for which a rough estimate gives a mass of $10^{-8}\mathfrak{M}_\odot$. The question is whether, in a medium of such very low density, turbulent motions will be produced. The answer depends upon the Reynolds number

$$R = \rho L v / \mu,$$

where ρ is the density, L is the thickness of the layer, v is the velocity of the gaseous stream, and μ is the viscosity. In the case of a typical ring we found $n_e = 10^{11}$. Since the ionization of H is responsible for nearly all the free electrons we have $n_H = 10^{11}$, approximately, and $\rho = n_H m_H = 10^{-13}$ gr/cm^3. The thickness of the ring is $L = 2R_* = 10R_\odot = 10^{12}$ cm. The velocity of the stream is the difference in the linear rotational velocity of the ring and the star's photosphere; hence, roughly, $v = 100$ km/sec $= 10^7$ cm/sec, and the viscosity of hydrogen is $\mu = 10^{-4}$. Inserting these values in the expression for the Reynolds number, we find

$$R = 10^{10}.$$

The theory of turbulence predicts that when $R > 10^3$ the stream motion is turbulent, when $R < 10^3$ it is laminar. We see that R is so large that turbulence must exist, even though the values of the parameters we have used are very uncertain.

Another important difference between the theoretical nebula and the one observed is its size. Von Weizsaecker assumes $r_{max} = 5 \times 10^{14}$ cm., while we find $L = 10^{12}$ cm. It is not yet possible to reconcile these discrepancies, but perhaps they are not sufficiently serious to destroy the analogy.

The existence of turbulence in our ring-like structures is an observed fact. Another fact is the absence of laminar flow; we have seen that in Zeta Tauri, Phi Persei, Pleione, and in other shells different layers show different rotational velocities. In the case of Zeta Tauri the observations give the following rotational velocities

He I (diffuse singlets)	$v_0 = 400$ km/sec
He I (diffuse triplets)	$v_s' = 350$
Mg II	$v_s' = 200$
Si II	$v_s' = 100$
Fe II	$v_s' = 75$
Ca II	$v_s' = 50$
He I 3965	$v_s' = 50$
H	$v_s' = 50$

Conservation of angular momentum would demand that

$$rv_s = Rv_0,$$

where v_0 is the rotational velocity of the star's reversing layer and R is its radius, while v_s is the rotational velocity of the ring and r is its radius. But the rotational broadening of the lines produced in the ring gives us, not v_s, but

$$v_s' = v_s \sin \theta = v_s \, R/r,$$

where θ is the angle subtended by the star's radius at any representative point in the ring. The geometrical dilution factor is

$$W = \frac{\omega}{4\pi} = \frac{R^2}{4r^2}.$$

Hence,

$$W = \frac{1}{4} \; ; \; v_s'/v_0.$$

Inserting the observed values for the reversing layer (v_0) and the outermost layer of the ring (v_s'), as given, for example, by the He I line λ 3965, which has a metastable lower level, we find

$$W = 0.03.$$

This is in good agreement with the independent determination, $W = 0.01$, obtained from the intensities of the He I line levels. The corresponding ratio of the radii is

$$R/r = 0.34.$$

The actual rotational velocity of the ring is

$$v_s = v_0 \, R/r = 140 \text{ km/sec.}$$

This may be compared to a value of $v_s = 200$ km/sec, which results from the overall width of the Hβ emission line. The agreement is satisfactory.

The conditions are favorable for the operation of von Weizsaecker's mechanism.

III

THE ORIGIN AND DEVELOPMENT
OF CLOSE DOUBLE STARS

1. Frequency of Binaries

IT HAS been estimated that at least one-fifth of all the stars are binary systems. In Figure 3 we demonstrated the distribution of 39 stars which are closer to the sun than 5 parsecs. Of these, 21 are believed to be single, 7 are double (but for two of these only one component could be plotted), and 2 are triple. It is clear that the existence of multiple stars must be regarded as a common occurrence among the building blocks of our galaxy. Somewhat similar results have been obtained in other ways. For example, from a discussion of the stars in the open cluster Praesepe, Heckmann and Haffner found that no less than twenty per cent of the members of this stellar association must be regarded as binaries. Their method consisted in examining an H-R diagram of the cluster. The great majority of the stars fall within a single narrow sequence. But a few stars depart from this narrow sequence in such a way as to suggest that what had been observed as a single star is really an unresolved binary.

An even more convincing discussion has been prepared by Kuiper. Figure 4 shows the H-R diagram for the stars which have parallaxes of 0.095″ and larger. Kuiper has used this material to study the frequency of binary stars. In this typical sample of the stellar population in the vicinity of the sun, he finds that fifty per cent of the objects of types A to K are either binaries or more complicated multiple stars. This does not account for observational selection, so that undoubtedly there are other objects which have not as yet been discovered as binaries. For example, G. Van Biesbroeck has been engaged for a number of years in the study of stars of large proper motion with the McDonald 82-inch reflector. His purpose is to discover distant companions which have the same proper motion as the known stars of large angular motion across the lines of vision. Several stars of this kind have recently been found. They are very wide binary systems, and because of their large proper motions they must be close to the sun. In an earlier paper, Kuiper had discussed the effect of the observational selection and had found that as many as eighty per cent of all the stars may be binaries or multiple stars. He has also made a study of the distribution of binary stars in relation to spectral type. In a general way, the ratio remains about the same—fifty per cent—except that the M dwarfs seem to have fewer binaries. But this is not yet certain because these stars are always faint and difficult to discover as binary systems. Luyten has discovered at least one system in which both components are white dwarfs. There are several other binaries in which one component is a white dwarf.

Kuiper has divided his material with regard to the total mass. In this manner he found that of those stars whose masses are larger than the mass of the sun, sixty-eight per cent are known to have divided into two or more parts. For masses which are smaller than the mass of the sun or equal to it, the fractions are considerably smaller. For example, for an interval between 1/2 and 1 solar mass, the fraction is about thirty per cent. For an interval of 1/4 to 1/2 the mass of the sun, the fraction is only fifteen per cent and, finally, for stars with total masses between 0.1 and 0.25 that of the sun, the fraction is only eight per cent. But, again, the observations of stars of small mass are probably somewhat incomplete.

The wide distribution of binaries in our galaxy is thus firmly established by all statistical investigations. We are undoubtedly concerned with a problem of major cosmogonical importance and we must attempt to explain why and how the binary stars have originated.

Aside from the photometric method of Heckmann and Haffner, which makes use of the H-R diagram, we are limited in the discovery of binary systems to three different procedures. First, there is the direct observation of stars through a telescope, resulting in the recognition of a large number of close doubles. Visual double stars have been known for a long time; but Sir William Herschel, with his large reflecting telescopes, was the first to undertake systematic surveys of all bright stars in order to discover whether any of them were binaries. This work was continued by many other observers, especially during the nineteenth and the early part of the twentieth centuries. The results of thousands of discoveries of visual systems have been incorporated in a general catalogue of double stars, by R. G. Aitken, which was published by the Carnegie Institution of Washington in 1932. Although the stars included in it are limited in apparent magnitude and in distance in such a way that for the fainter stars only relatively narrow pairs were included, the catalogue contains a total of 17,180 stars between the limits of the North Pole and declination $-30°$. The southern double stars were catalogued by R. T. A. Innes. Extrapolating this number over the entire sphere, we would be justified in concluding that there exist approximately 23,000 binaries of the kind included by Aitken. Down to the ninth apparent magnitude, there is about one visual binary for every eighteen stars. This is a surprisingly large number if we consider that among the brighter, and especially the fainter, stars, only a small fraction are sufficiently close to us to show binary structure, unless the two components are exceedingly far removed from one another.

It can be shown from statistical considerations that only a small fraction of the visual binary stars can be due to the chance arrangement of the stars in the sky. Binaries which are the result of chance are called optical pairs. They are so rare that they can be neglected in practice. Those systems which demonstrate binary motion are especially interesting to us, but at present only a small fraction show a change in the relative positions of the two components. An even smaller number have been sufficiently well observed for the determination of their orbits. According to recent estimates by W. H. Van den Bos, W. S. Finsen, and W. Rabe, there are about 250 systems for which orbits have been computed. The longest period of revolution thus far determined is approximately 11,000 years, in the case of the binary Sigma-two Ursae Majoris. This may be contrasted with the system BD −30°181, whose period is only 4.6 years. An interesting pair which has not been observed with the ordinary micrometer, but which has been followed by means of an interferometer, is Capella, or Alpha Aurigae. Its period is 104 days. This star forms a transition between the visual systems and those which we describe as spectroscopic binaries.

The visual binary of the shortest known period is BD −8°4352, which was found by Kuiper to have a period of only one year and eight months. This period is comparable to the longer periods of the spectroscopic binaries. In fact, there are many of the latter which are a great deal longer; for example, the well-known spectroscopic double star Epsilon Aurigae has a period of twenty-seven years. Despite this long period, it is so far away that it is not visible as a binary through the telescope, but can be detected only by the Doppler effect in its spectrum.

The second method of discovery is provided by the Doppler effect and the use of a stellar spectrograph. In 1889 at the Harvard Observatory, E. C. Pickering found that the bright star Zeta Ursae Majoris, or Mizar, which was then already known as a visual pair, has a brighter component whose spectrum varies in a period of about 20½ days.

On certain nights the spectrum of this star appeared to have the usual kind of absorption lines, which resemble the lines of the solar spectrum. On other nights all lines appeared as close pairs. Pickering soon realized that he was concerned with a binary system consisting of two stars revolving around each other in a period of 20½ days. When the two components are moving across the line of sight, they produce spectral lines without a Doppler shift, so that the two sets of lines from the two components exactly coincide. But when the two stars are oriented in such a way that one is receding from us while the other is approaching, the spectral lines are affected by their Doppler displacements. The lines of the receding component are shifted toward the red side of the spectrum, and the lines of the approaching component are shifted toward the violet side of the spectrum. The velocity of the motion of each component is sufficient to separate the two sets of lines completely.

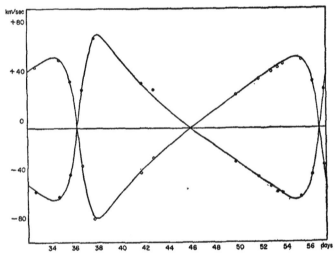

FIGURE 21. Velocity-curves of the spectroscopic binary Zeta-one Ursae Majoris (Mizar).

From the measurement of such pairs of lines it is possible to determine the orbital velocities of the two components, and this, in turn, leads to the construction of a velocity-curve for each member of the binary system, representing the manner in which the component of the star's motion projected upon the line of sight varies as a function of the time. Figure 21 shows the results of the measurements of Zeta-one Ursae Majoris made by C. U. Cesco. The velocity-curve of each component is highly unsymmetrical. If the orbits of the two components had been circular, the resulting velocity-curves would have been simple sine curves. Instead, they are steep on one side and less steep on the other. This lack of symmetry is the result of eccentric motion. From its measurement we can determine the eccentricity of the orbit and the orientation of the longer axis of the elliptical orbit. The spectra are shown in Plate XIV. It is of interest that F. G. Pease succeeded in 1925 and 1927 in measuring the two components of this spectroscopic binary as a visual double star with the help of a 20-foot interferometer attached to the 100-inch telescope of the Mount Wilson Observatory. The visual orbit was computed by Russell, who found that the apparent semi-major axis is 0.01″.

A catalogue of spectroscopic binary systems, by J. H. Moore of the Lick Observatory, in 1949, lists 524 orbits. Several hundred additional binaries have been discovered because of changes in the measured radial velocities, but their orbits have not yet been determined. At one time it was thought that perhaps one out of every three or four stars of early spectral type, that is, those having a surface temperature between 50,000° and 20,000°, is a spectroscopic binary. It is possible that this estimate is too large, but in all probability a considerable fraction—certainly not less than twenty or twenty-five per cent—of the stars of high surface temperature are spectroscopic binaries.

In the case of the visual binaries we had seen that observational selection favored the discovery of systems which are relatively close to the sun and in which the distances between the components are large. In the case of the spectroscopic binaries there exists a similar observational selection. We discover primarily those systems which have large masses. That makes for a large orbital velocity. Moreover, we discover more readily those systems whose orbits are oriented approximately in the line of sight. The reason is that the Doppler effect applies only to motion in the line of sight, that is, motion of approach or recession. It does not register motion across the line of sight. Hence, if a star is moving at an angle to us, all we can observe in the form of a displaced spectral line is that component of the motion which registers approach or recession. Accordingly, binary systems whose orbits are oriented at right angles to the line of our vision show no effect. Their components are constantly moving at right angles to the line of sight and their lines will always be blended so that no duplicities will appear on our photographic plates. Finally, there is one further selection which is of importance in connection with the discovery of spectroscopic binaries. There is a definite preference for the discovery of those systems whose orbits are eccentric. The reason for this is that eccentricity tends to alter the velocity in such a way that it is large at some points of the orbit and relatively small at the opposite point. When the two stars pass through the points of closest approach, which we describe as periastron passage, the orbital velocity is large. At the opposite point of the orbits, which we describe as apastron passage, the velocity is small. It is for this reason that we can occasionally discover a binary system of large eccentricity when we might not be able to discover the separation of the lines in a circular orbit produced by components of exactly the same mass.

We cannot always measure the components of both stars in the spectrum. We do so only when the two luminosities are nearly the same. In the case of Mizar this is true, but there are many spectroscopic binaries in which only one component is bright enough to register its light on our photographic film. Generally speaking, the fainter component must differ by not more than one stellar magnitude from the brightness of the principal component in order to have its spectrum registered on the photographic plate. It turns out that this is rarely the case. Hence the majority of the orbits which have been measured rest upon velocity-curves of only a single component. These single-line binaries are of great interest, but they do not provide us with as much information as the double-lined systems. In particular, we lack information concerning the masses of the stars. When two components have been observed, we determine separately the masses of each component multiplied by the cube of the sine of the orbital inclination. This latter factor, depending upon the inclination, cannot be determined from spectroscopic observations alone. In the case of a single-lined spectroscopic binary we derive only a more complicated function of the mass of that component, the mass ratio of the brighter component to the fainter component, and the cube of the sine of the inclination. This quantity is described as the mass function of a spectroscopic binary (see page 27). If the two components were of the same mass, which of course we can never be absolutely sure of unless we have observed the two components in the spectrum, then the mass-function gives us exactly one-quarter of the mass of the fainter component, which is also equal to the mass of the brighter component, each multiplied by the cube of the sine of the inclination.

The third and in some respects most important method of discovering binary systems rests upon the observation of eclipses. Since there are so many binary systems within our galaxy, it is not surprising that a considerable number are oriented in such a way that their orbits are almost exactly in the line of sight. When this condition is fulfilled, then once in every cycle the brighter component eclipses the fainter component, and once in every cycle the fainter component eclipses the brighter component. These eclipses produce marked variations in the brightnesses of the stars. A famous example is the Demon Star, or Algol, also known as Beta Persei. It must have been known as a variable for a long time, although its recorded discovery was made only in 1783 by the English astronomer Goodricke, who wrote concerning it, "I should imagine that it could hardly be accounted for otherwise than . . . the interposition of a large body revolving about Algol." Accurate observations of the brightnesses of such eclipsing variables yield what is called a light-curve. H. N.

Russell of Princeton University has developed the theory of the determination of the orbital elements from the properties of the light-curves. The first complete discussion of 87 eclipsing systems, based upon Russell's method, was made in 1913 at Princeton by Harlow Shapley. A compilation by S. Gaposchkin in 1938 listed 268 orbits. An even more recent "finding list" of eclipsing variables in 1947 by N. L. Pierce gives 546 such systems. Not all of them have accurately determined orbits. The 1948 General Catalogue of Variable Stars by B. V. Kukarkin and P. P. Parenago contains 1,913 eclipsing stars.

As in the case of the visual binaries and the spectroscopic binaries, there are important observational selections which limit or restrict the discovery of binary stars from their eclipses. First, there is the limitation of inclination. We can find only those stars in which eclipses really take place. This statement must be slightly qualified. It is known that the binary systems in which the two components are close to one another show deformations in the shapes of the two stars. Instead of being spherical or spheroidal they become elongated, almost egg-shaped, pointing their longer axes toward each other. These ellipsoidal shapes of stars produce a variation of the light of each component, even though there may not be a real eclipse of one component by the other. What we observe is simply the gradually changing cross section of the star as seen from the earth. When the two components are side by side they are pointing their longer axes toward each other and expose to our view their greatest cross sections. On the other hand, when the two stars are in conjunction, they are exposing to our view their smallest cross sections. Hence, their light is least at these latter times, and greatest at the former. A few systems of this ellipsoidal type are known. In many others the ellipsoidal variation in light is superposed over that produced by real eclipses. In both, the probability of discovery is greatly increased when the inclination is 90°, and it is negligible when the inclination is smaller than about 70°. Next, the discovery of an eclipsing system is increased in probability if the conditions are such as to favor the production of deep eclipses, or large light losses.

A deep eclipse is produced when a star of relatively small surface brightness totally eclipses another star of large surface brightness. Hence we favor the discovery of binary systems in which the two components are very different in surface brightness. Such a system, for example, is U Cephei. Its principal star is one of spectral type B, which corresponds to a temperature of perhaps 15,000°. The surface brightness of this star is very great. Its companion is a star considerably larger in size but having a much smaller surface brightness and a surface temperature of about 6000°.

During the eclipse, when the feebly luminous, large companion obscures the light from the small bright primary, we can make the exposure times at the spectrograph long enough to record the light of this companion. It turns out to be a star of spectral type approximately G, like the sun, but having characteristics which we normally associate with the group of stars described as subgiants. These stars are not real giants like Alpha Persei, or Capella, but they are somewhat less luminous—between the ordinary main-sequence stars and the true giants. They are larger in size than the main-sequence stars and they show certain line intensities which we associate with fairly high luminosity. It is interesting to note that in the majority of binaries of this kind, which we describe as typical Algols, the secondary components have spectra of the subgiant class. Ordinarily, among single stars subgiants are not frequent. We observe either main-sequence stars or giants, and only very occasionally a subgiant. In the eclipsing systems, subgiants are definitely more frequent among the secondary components than are either main-sequence stars or true giants or supergiants, as in Epsilon Aurigae, VV Cephei, and Zeta Aurigae.

Observational selection also favors the discovery of eclipsing variables of short period. It is even more important in this group than in the spectroscopic binaries. The importance of observational selection is very great and it affects the interpretation of the observational results. If all binary systems were alike in their physical characteristics, then it would make no difference whether we discovered them by the visual, the spectroscopic, or the photometric method. We would simply conclude that if we had made the discovery photometrically, the inclination must be close to 90°. Discovery of a system by means of a spectroscope would not place quite so stringent a restriction upon the inclination, but we would still be inclined to believe that the inclination is not very far from 90°, though perhaps not sufficiently so to produce eclipses. If we had discovered a system by the visual method, we should conclude that the inclination could be anywhere from zero to 90°. But the physical properties of the binary systems are not all the same. Accordingly, by means of each method we discover groups of stars which differ greatly from one another. As an example, let us consider the relation between the orbital velocity of a binary system and its period. It is clear that such a relation should exist, because it is demanded by Kepler's third law. When we plot the velocity of the bright component of the system, or, what is almost the same, the semi-amplitude of the velocity-curve, K_1, (corrected for orbital eccentricity) against the logarithm of the period, we obtain for the B and A stars diagrams shown in Figure 22. We notice that the stars which are shown by the black dots all fall within a roughly triangular area.

All these binaries were discovered by their Doppler displacements. But if we plot those systems whose brighter components are also of classes B and A, and which were discovered from their eclipses, we obtain Figure 23, in which the limiting curves were drawn from Figure 22. We have already discussed the significance of these diagrams in connection with the determination of the masses of the stars (p. 30). What interests us here is the fact that observational selection has resulted in two entirely different distributions. We should have expected

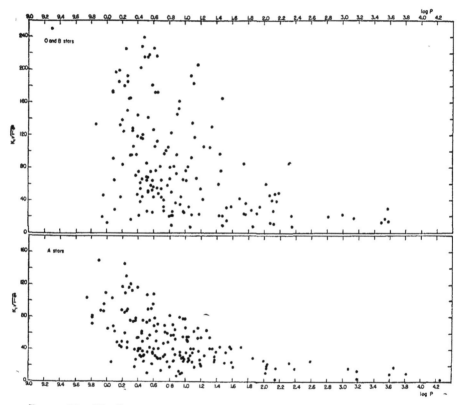

FIGURE 22.   Distribution of spectroscopic binaries of classes O, B and A according to K_1 and P.

that, because in the second group $i = 90°$, the points would all lie close to the upper limiting curve, whose equation, by Kepler's third law is $K_1 = CP^{-1/3}$, where C is a constant depending upon $\mathfrak{M}_1$ and $\mathfrak{M}_2$. The stars discovered by means of a spectrograph are preferentially objects of large mass and also objects in which the mass ratio is approximately equal to one. This is due to the fact that if the mass ratio were very different from one and we observe only a single spectrum, the more massive and more luminous component would show

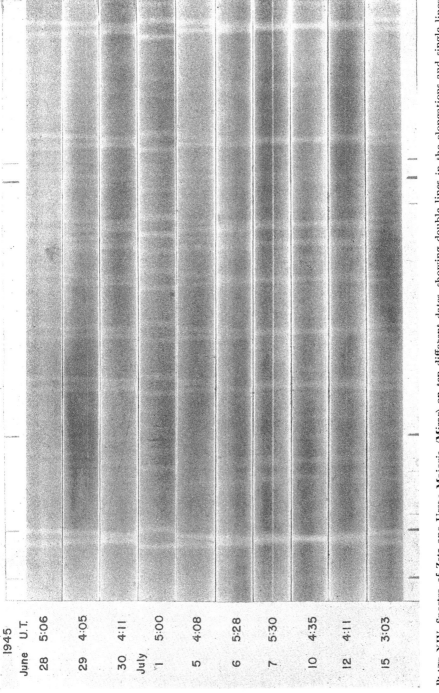

PLATE XIV. Spectra of Zeta-one Ursae Majoris (Mizar) on ten different dates, showing double lines in the elongations and single lines in the conjunctions.

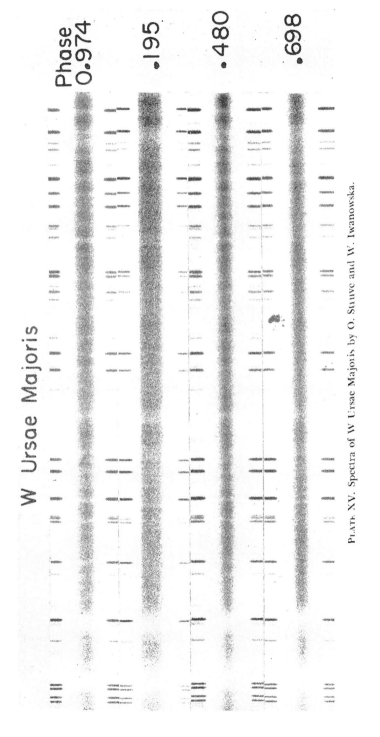

W Ursae Majoris

Phase
0.974

.195

.480

.698

PLATE XV. Spectra of W Ursae Majoris by O. Struve and W. Iwanowska.

a smaller Doppler displacement than it would if the value of $\alpha = \mathfrak{M}_1/\mathfrak{M}_2 = 1$.

It would be advantageous for the discovery of binary systems if the brighter component, whose spectrum we can record, should be the one of smaller mass, because it is the less massive component which has the larger orbital velocity. Such a group of systems exists in the galaxy, but it must be anomalous, since it would violate the usual

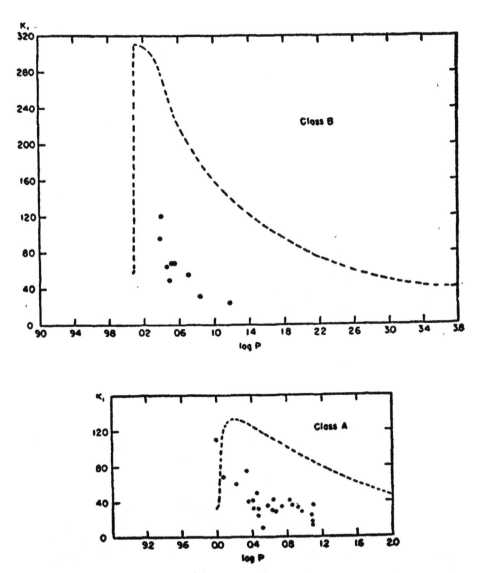

FIGURE 23.   Distribution of eclipsing variables of classes O, B and A according to K_1 and P.

mass-luminosity relation. The exception occurs in the case of the Wolf-Rayet binaries and a few other systems. A number of binaries have one component whose spectrum is of the Wolf-Rayet type, with strong emission bands of great width; the other component is usually an ordinary O-type or B-type star with absorption lines, corresponding to a temperature of 20,000° or 30,000°. A famous example of this kind is V 444 Cygni, which has been the subject of numerous investigations at the Mount Wilson, Lick, McDonald, and Harvard Observatories. In all of these binaries the Wolf-Rayet component, which is the stronger in the spectrum, is the less massive and is therefore the one which gives the velocity-curve of larger range.

But these cases are exceptions. In ordinary binaries with spectra that do not show peculiar features, the components usually obey, at least approximately, the mass-luminosity relation.

In each group there must be a certain admixture of stars of the other kind. For example, among the stars discovered by their Doppler effects there must be a certain number which are similar to the ordinary eclipsing variables. A diagram prepared some years ago by J. Stebbins shows precisely this effect. He expected to find all eclipsing variables near the upper limiting curve of the diagram. In reality, the stars near the upper limiting curve were almost all eclipsing variables, but the reverse was not true. He found some eclipsing variables which were not near the upper limiting curve but fell far below it. These eclipsing variables are typical of the group included in Figure 23, even though they had been discovered by means of the spectroscope. It would be a tempting statistical task to disentangle the mixture of objects in these two groups. Among the stars discovered photometrically, there must also be a certain number in which the mass ratio is one or nearly one, and the problem is to try to disentangle the distribution of the objects of different mass ratios.

2. The Ellipticity Effect

We are now coming to the discussion of a different feature of the spectroscopic binaries, as well as of the eclipsing variables. Among the latter it has long been known that an important property is the ellipticity in shape of the stellar components: the two stars are not spherical, but more nearly resemble elongated egg-shaped bodies. The gradual revolution of such elongated bodies around their common center of gravity produces a continuous variation in light, even at those phases where there is no obstruction of light by an eclipse. This effect may therefore be regarded as a measure of the degree of closeness of the two components. Figure 24 shows a diagram first prepared by D. J. Martynov in the Soviet Union. The degree of

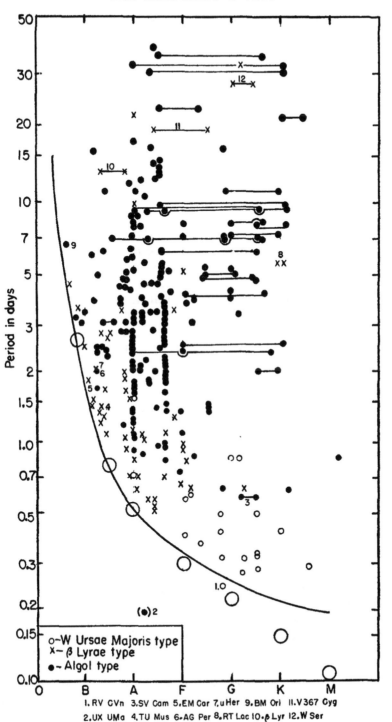

FIGURE 24. Distribution of different kinds of eclipsing variables according to period and spectral type.

ellipticity is closely correlated with the period, in eclipsing variables. This is well shown in a table by W. Becker:

TABLE X

Ellipticity in Eclipsing Variables

Period	Without Ellipticity	With Ellipticity
$<1^d$	86	162
1–5	477	68
5–10	108	10
>10	79	14

Martynov's diagram shows that nearly all points fall above a clearly defined limiting curve. The large circles represent the theoretical relation between P and T (or spectrum) if the two components are in contact and if they have average radii for each spectral class. The fact that this curve agrees well with the observed distribution shows that among the known eclipsing variables systems with lower-than-normal luminosity (such as UX Ursae Majoris) are very rare. Immediately above the curve we see almost entirely small circles and crosses, that is, systems whose light varies throughout their cycles. These stars, of the Beta Lyrae and W Ursae Majoris types, are the ones which show ellipticity. Lying nearest to the limiting curve, they are the ones in which the components are almost in contact.

3. Axial Rotation of Close Binaries

We shall now consider a different property of these two groups of binary systems. Not only do they have a large amount of ellipticity and a pronounced reflection effect—a phenomenon well known to the photometric observers—but they are also characterized by large values of their rotational velocities. As a matter of fact, the existence of rotational motion in stars was definitely established because of the diffuse appearance of spectral lines in the system of W Ursae Majoris. W. S. Adams and A. H. Joy at Mount Wilson investigated the orbit of this short-period eclipsing system and found that the two components rotate with a period similar to that of their orbital revolution. This rapid rotation of each component causes the spectral lines to be diffuse.

Since then, many other stars with diffuse spectral lines have been discovered and in the majority of cases the diffuse contours have been attributed to rotational motion. All spectroscopic binaries and eclipsing variables of short period have more or less diffuse spectral lines. Thus there can be no doubt that their components rotate rapidly. Several investigations have been made to find whether the orbital periods and the periods of axial rotation of the two components are identical. It was first found by P. Swings and O. Struve that in the majority of spectroscopic binaries of short period there is equality of the two periods. But in systems of relatively long period—longer than four or five days—this equality breaks down, and we frequently observe components which rotate more rapidly than they would if there were synchronism of the two periods. There are, however, some exceptions even among the systems of relatively short period. The best-known example of this kind is U Cephei. Its orbital period is 2.49 days, and the rotational velocity at the equator of the brighter, smaller, and hotter component is 200 km/sec. If the periods of rotation and revolution were equal, we should obtain for the radius of the brighter, B-type, component

$$r_B = 7 \times 10^6 \text{ km},$$

or approximately ten times the radius of the sun. But from the orbital elements of the system we know that the semi-major axis of the orbit of the brighter component around the center of gravity of the system, multiplied by the sine of the inclination, is equal to

$$a_B \sin i = 4 \times 10^6 \text{ km}.$$

We have no accurate determination of the mass ratio of this system, but observations have been made at Mount Wilson and at McDonald of the fainter component during the total eclipse. Although this eclipse does not last very long, it yields a small portion of the velocity-curve of the fainter G-type component, and from the slope of this section, compared to the slope of the velocity-curve of the brighter component, we can derive an approximate value of $\alpha = \mathfrak{M}_1/\mathfrak{M}_2$, and this turns out to be about 1.5, or at most 2.0. Hence, the semi-major axis of the orbit of the fainter component $a_F \sin i$ is approximately equal to 8×10^6 km.

The photometric results give us the ratio of the radii of the two stars, $r_B/r_G = k = 0.62$, and we can conclude that $r_G = 11 \times 10^6$ km. Thus, we find that the sum of the radii of the two stars would be about 18×10^6 km, or about fifty per cent more than the sum $(a_B + a_F)$ $\sin i$. The inclination is so nearly 90° that we can neglect the quantity $\sin i$. Since this is impossible, we must conclude that the period of rotation is less than the period in the orbit. This conclusion is supported by the observational result that the absorption lines of the large cool secondary component of U Cephei are sharp and give a velocity of rotation of not more than about 50 km/sec. It is probable that this secondary component does rotate in synchronism with the orbital revolution. The brighter component rotates very much too fast.

An important result of the spectrographic observations which has not yet been entirely explained, but which is shared by the eclipsing variable RZ Scuti, is that the rotational velocities as determined by the hydrogen lines and by the helium lines are not the same. The helium lines give the large rotational velocities which we have used here. The hydrogen lines give smaller rotational velocities by a factor of two or three. It is difficult to explain this difference. At one time it seemed that we could suppose the hydrogen to form a stratum, or shell, high above the rest of the reversing layer of the brighter components of U Cephei and RZ Scuti, which does not share the rapid rotation of the star itself. This explanation is logical and it may turn out to be in accordance with the facts, but it is surprising and somewhat disquieting that we cannot definitely rely upon the contours of the spectral absorption lines to yield the actual velocity of rotation of the star itself. If we can have different rotational velocities at levels which are relatively close together, then what can we conclude concerning the rotation of the main mass of any star?

The interpretation of the very large rotational disturbances in the velocity-curves of U Cephei and RZ Scuti presents one of the most fascinating problems of astrophysics. It should be pointed out at once that these large disturbances are not always present. For example, in Beta Lyrae and in Algol the disturbances are much smaller in amount and have been interpreted to mean that the velocities of rotation of the brighter components of these two systems are compatible with the idea of synchronism between rotational and revolutional periods. Why U Cephei and RZ Scuti, two systems with very different periods, should be so different from Beta Persei and Beta Lyrae is not known. This unusual difference makes it somewhat less convincing to say that the rapid rotation of the brighter component in U Cephei and in RZ Scuti may have been produced by their evolution. One might have thought, for example, that the brighter component had contracted rapidly after the binary had been formed, and that the conservation of rotational momentum had caused it to rotate more rapidly than the larger component, which had remained in the subgiant state. It is also possible that during the formation of the binary the more massive component retained more than its normal share of angular momentum. This is supported by the fact that in systems with $\alpha = \mathfrak{M}_1/\mathfrak{M}_2 = 1$ there is almost never such a difference in rotational period.

4. Classification of Eclipsing Binaries

Before we can make further progress in studying the evolution of double stars we must classify them according to their physical properties.

There are a bewildering number of different kinds of eclipsing variables. The usual classification of these stars is based primarily upon observational features. We distinguish between stars of the Algol type, of the Beta Lyrae type, and of the W Ursae Majoris type. But it has been pointed out by various investigators that this simple classification into three groups is not sufficiently refined. It does not subdivide the Algol stars and the Beta Lyrae stars into physically significant groups. We shall follow the procedure adopted by the Russian astronomer T. V. Krat. He introduces seven different classes, described as: A, B, C, D, E, F, and G. Class A consists of systems both components of which belong to spectral types O to B9. The type star is Beta Lyrae, but UW(29) Canis Majoris, AO Cassiopeiae and others would be better examples. The radii of these components are usually comparable with the semi-major axes of their orbits. Hence, these systems usually show a pronounced ellipticity effect. It is possible to subdivide this group still further. For example, there is a distinct subgroup which has relatively smaller stars, and consequently a less pronounced ellipticity effect. Typical for this subgroup are the eclipsing systems Y Cygni and VV Orionis. The orbital period is usually of the order of a few days.

Systems belonging to group B have fainter components belonging to spectral classes A0 to F5. The stars of this class have different periods, ranging from a few tenths of a day for stars like CG Cygni to several days for stars like Lambda Tauri.

In group C at least one of the components is a subgiant. The light-curves frequently show a constant phase at maximum, and in consequence the ellipticity of these stars is usually small. The periods are usually of the order of one day or longer, up to ten or more days. One of the systems of this group is X Trianguli, whose period is less than one day. Other representatives with longer periods are Algol, or Beta Persei, U Cephei, and Delta Librae. It is important to realize that a large number of eclipsing variables belong to this group.

In group D both components belong to the main sequence. The spectrum of the companion is later than F5. These systems are therefore composed of dwarfs. The periods range from a fraction of a day to several days. One of the systems of relatively long period is Alpha Coronae Borealis. The effect of ellipticity is small. Typical representatives are R Canis Majoris and RT Persei.

Group E consists of pairs which are similar to W Ursae Majoris. The period is always less than one day, and the spectral types of the two components are between F5 and M. W Ursae Majoris itself is the most typical representative. Another typical star is i Bootis B.

In group F, one or both stars are "degenerate." The only system known at the present time is UX Ursae Majoris.

Finally, we come to class G. Both components of these systems are giants or supergiants. The periods are usually several hundred days or longer. Typical representatives are Epsilon Aurigae, Zeta Aurigae, and VV Cephei.

It is useful to consider the populations of the different groups. Krat finds that there are ten stars of group A, twenty of group B, thirty of group C, eleven of group D, seven of group E, one of group F, and three of group G. In groups D, E, and F the periods are usually less than one day. In groups A, B, and C the maximum concentration is between periods of one to four days. In group G the periods are usually very long.

A great deal of work has been done in recent years towards amassing a large amount of observational material concerning the properties of these different groups of systems. It is useful to realize that the rapidly revolving systems, such as those of the W Ursae Majoris group, are always characterized by broad and diffuse spectral lines of the kind which we have described in terms of rapid rotation around an axis. New material on the W Ursae Majoris stars has been secured in 1943–1949 at the McDonald Observatory. All these systems have broad and diffuse lines and in all of them the axial rotation is rapid—of the order of 100 km/sec at the equator of each star. There is every reason to suppose that the axial rotation has a period identical with that of the orbital revolution.

It is, of course, not unusual to find eclipsing variables with broad spectral lines. All systems of short period have them. This was one of the strongest arguments in favor of the rotational interpretation of these lines. It is strange that the single stars of the same spectral types as the W Ursae Majoris binaries have sharp lines, and never show rapid rotation.

The eclipsing variables of Krat's groups A, B, C, and D also frequently show broad lines, although as a rule the spectral lines of the components of even those systems whose periods are less than one day are not as conspicuously broad as are the lines of the components of the W Ursae Majoris systems.

Particularly interesting is the fact that the single stars of the spectral types corresponding to Krat's classes A, B, C, and D are often characterized by very rapid axial rotations. We have already seen that, for example, among the B-type and the A-type stars which are not spectroscopic binaries, a large fraction, perhaps more than one-half, have rotational velocities of the order of 100 km/sec if measured at the equator. Every spectroscopist knows that it is difficult to find a B-type star with perfectly sharp lines, and among the brighter objects of this kind there are only a few which are adapted to the critical study of stellar absorption lines. All the rest have diffuse lines which are difficult to measure because their central absorptions are exceedingly small.

This difference in the relative rotations of binaries and single stars in groups A, B, C, and D on one side, and group E on the other side, constitutes one of the most significant facts in astrophysics and we shall deal with it in this chapter at some length.

5. The Two Principal Groups of Binaries

Our principal purpose is to discuss the problem of the origin and evolution of the double stars. We wonder why there are so many binary systems, how they have been produced, and how they evolve. It is unreasonable to suppose that all the different kinds of double stars which we recognize at the present time have been formed in their present condition. It is more reasonable to suppose that we observe now different stages in the evolution of double stars. But caution must be exercised. It would, for example, also be unreasonable to go to the other extreme and suppose that all binary stars were produced in exactly the same condition. Hence, the distribution of the observed physical parameters of binary stars represents the consequences of evolution as well as of the differences at the time of their origin. This makes the problem a difficult one, and we must use restraint in our interpretation of the observed facts. We shall discuss in detail two large groups, both of which may be described as contact binaries. In the first place there is a group of such binaries among the stars of early spectral type. Beta Lyrae is a famous example, although its spectrum and to some extent its light-curve are exceedingly complex and present many unusual features not found in other systems. Hence, a system like AO Cassiopeiae is a better representative. On the other side, there are the stars of the W Ursae Majoris type. They are formed of components of relatively late spectral class (F, G, and K), but these components are also close together, so that ellipticity of shape is one of their outstanding characteristics.

When we were discussing the ellipticity effect as a function of the period, we combined all those stars whose light-curves showed a change in brightness between the minima. Thus, we made no distinction between systems like AO Cassiopeiae, of type O, and W Ursae Majoris, of type F. But these stars are in many respects very different. AO Cassiopeiae consists of two stars of early spectral type, both O8. The masses are approximately 31 and 29 times the mass of the sun, respectively; the diameters are about 16 and 10 times the diameter of the sun; and the densities are approximately 0.01 and 0.03 of the density of the sun. The two luminosities are roughly 300,000 times that of the sun for the brighter component, and 100,000 times the brightness of the sun for the fainter component. The bolometric absolute magnitude of the brighter star is about -9; that of the fainter star, -8. The separation between the components is approximately 25 million kilometers, or 36 times the radius of the sun. Remembering that the sum of the radii of the two stars is 26 times the radius of the sun, it is clear that, even disregarding the ellipticity of the two components, the separation between their surfaces can only be of the order of 10 times the radius of the sun, or a small fraction of the total size of the system. Considering, further, that the stars must actually be elliptical in shape, as is shown by the curvature of the light-curve between minima, we must conclude that the surfaces are even closer together and that they may even touch.

Let us consider, next, a typical star of the W Ursae Majoris type. Shapley has recently summarized our information concerning their light-curves. He finds that on the average these stars, which he designates by the letter W, belong to spectral class F8. The average period is 0.49 day. The range of the light-curve at primary minimum is 0.65 magnitude, and the range at secondary minimum is 0.60 magnitude. Thus the two minima are almost of the same depth. Of course there are individual differences. In some binaries the minima are so nearly alike that it is difficult to decide which is the primary. In others the minima are not quite so similar. The average orbital inclination is 74° and the ratio of the radii of the two components, as determined from the light-curve, is 0.88. Finally, the average ellipticity of the components is 0.76. It is sometimes difficult to distinguish from the light-curves alone whether one is dealing with a Beta Lyrae-type star or with a star of the W Ursae Majoris class. Shapley, in his recent work, somewhat arbitrarily postulated as a working definition of the W Ursae Majoris class of binaries that the period should be less than one day. The spectrum should be later than A0. The unrectified range at secondary minimum should be greater than 2/3. The unrectified range at primary minimum and the light-curve between the two minima must show conspicuous ellipticity variations. The restriction that the spectral type must be later than A0 eliminates all binaries whose principal component is of spectral class B or O.

It is of interest that the very great majority of W Ursae Majoris variables are confined in spectral type to a relatively narrow range. They seem to cluster around late G or early K. They invariably show characteristics which we usually associate with spectra of stars of the main sequence. There are no giants among them. Measures of the parallax suggest a photographic absolute magnitude of +4.9 at maximum. Shapley believes that +5.0 is a slightly better estimate. The dispersion in absolute magnitude is small—so small, in fact, that one could easily use these stars for the determination of distances.

In the other group of systems, the spectral types range from O to M and the absolute magnitudes from −3 and even brighter to +9 for the faint distant companion of Castor, which is also called YY Geminorum. Shapley has designated this group by the letter A, although their spectra range from B5 to A8 and their absolute magnitudes from −1 to +3. The few systems outside these ranges are relatively infrequent; hence, for practical purposes, we may follow this grouping. Although the spread in absolute magnitude among the A systems is quite large, we shall use as a mean value +1.0.

Shapley has made a study of the frequency distribution of the binaries of groups A and W. The latter group includes all W Ursae Majoris stars which satisfy the conditions he had set for them. In the catalogues of eclipsing variables the great majority of systems belong to group A. For example, H. Schneller's catalogue for 1940 has approximately 930 systems of class A and 150 systems of class W. If the space-distribution of the binaries of groups A and W were the same, then in view of the great difference in their absolute magnitudes we should expect to observe a much greater number of A stars than W stars, provided our search is complete to a certain apparent magnitude. The reason for this is that for systems of absolute magnitude +1 we are able to penetrate much deeper into space than for systems whose absolute magnitude is +4 or +5. A simple computation of the volumes involved shows that the ratio is approximately 125. In other words, if we had a uniform distribution throughout space of A systems and W systems, we should observe 125 systems of group A for every system of group W. Shapley pointed out that Schneller's catalogue indicates that the W stars are much more frequent in space than can be reconciled with the assumption of uniformity in distribution, and that they are more frequent than the A stars. He finds that on the average the W stars are approximately 20 times more numerous per unit volume than are the A stars.

This result is confirmed by the distribution of the A and W systems in different galactic latitudes. A large survey at the Harvard Observatory in which counts of systems were made to a limiting magnitude of 14.0 or 14.5 gave the following results: If we designate by R the volume ratio dependent upon the absolute magnitudes adopted for the two systems, +1.0 for the A systems and +4.5 for the W systems, and if we next designate by W/A the ratio of the number of W binaries to the number of A binaries, then we can write $F = R \times W/A$. The quantity F is a function of the galactic latitude. For example, within the limits of 55° to 90° northern and southern galactic latitude, Shapley finds $F = 115$. For limits of 35° to 45° in galactic latitude, $F = 57$. For limits 0° to 12.5°, $F = 15$. This change in the value of F with the distance from the central plane of our galaxy is caused by the fact that for the A systems the Harvard search has penetrated to distances of about 4,000 parsecs, while for the intrinsically faint W systems it has penetrated to only 800 parsecs. In the galactic plane and near it, the space densities of the A and W systems remain approximately constant, so that F has its normal value of about 20. But at right angles to the galactic plane the general density of all stars decreases rapidly with the distance, so that at 4,000 parsecs all stars, including the A systems, are very rare per cubic parsec, while at 800 parsecs the star density is still large. Hence F becomes very large in regions close to the galactic pole.

Shapley concludes that the eclipsing variables of the W Ursae Majoris class are more numerous in space than all other variable stars taken together. This result is of exceptional interest if we recall that at the same time the single stars of spectral classes A and B are much less numerous in space than are the stars of classes G and K. For example, if we examine the H-R diagram in Figure 4, which was constructed for the nearby stars, we find that the ratio of stars within our neighborhood whose spectral types are B and A to the number of stars of later spectral types is about 1 to 20. Thus, it seems probable that Shapley's result simply indicates that the W binaries are more numerous than the A binaries by about the same factor by which the single stars of late spectral class in the main sequence are more numerous than the single stars of spectral classes B and A.

We do not know exactly the ratio of single stars of any given spectral subdivision, such as A0 or K0, to the number of binaries of groups A and W, respectively. From the preceding discussion it appears that this ratio should be about the same. As a rough estimate we may adopt:

$$100 < \frac{\text{Number of single stars of a spectral subdivision (B or A)}}{\text{Number of A-group binaries}} < 1{,}000$$

$$100 < \frac{\text{Number of single stars of a spectral subdivision (dwarfs F, G, or early K)}}{\text{Number of W-group binaries}} < 1{,}000$$

But, among stars of spectral class later than K5, the corresponding ratio is probably different:

$$\frac{\text{Number of single stars of a spectral subdivision (late dwarfs K or M)}}{\text{Number of binaries with ellipticity}} \gg 1{,}000$$

This latter conclusion is based upon rather inadequate information, because the K and M dwarfs are intrinsically faint and few have been studied for light variation. But YY Geminorum (Castor C) of spectral class M is not a binary which resembles W Ursae Majoris in ellipticity. While we are not yet certain, we suspect that the evolutionary history of a close binary usually precludes it from becoming a pair of M dwarfs in close contact, like W Ursae Majoris.

We are more certain that, in discussing the evolution of close double stars, we must give at least as much consideration to the little-known W systems as to the more spectacular and more fully investigated A systems.

6. The Spectra of the W Ursae Majoris Binaries

These stars are faint and it is not easy to secure adequate spectroscopic material. Only eighteen of these stars are now known which are brighter than the tenth apparent magnitude, and the brightest are fainter than magnitude 6.5, at maximum. Only during the past few years telescopes of large size have been employed for their study. At the Mount Wilson and the McDonald Observatories ten systems have now been analyzed by means of spectrographs of sufficient dispersion. As a rule, the spectral types are G or early K. Usually both components are visible in the spectrum, and the relative intensities of the lines differ but slightly. The more massive component is eclipsed during the secondary minimum, even though in some cases there may be doubt about the selection of the minimum which is considered as primary. This is illustrated by the observations of AH Virginis by Y. C. Chang at the McDonald Observatory. Figure 25 represents the velocity-curves obtained by him. The radial velocity of the more massive component, which has the smaller orbital velocity and therefore the smaller range in the diagram, changes from negative values to positive values at zero phase. This phase was chosen to coincide with the deeper of the two minima in the light-curve. In this system the difference is quite pronounced. The minimum at primary eclipse is 10.74 magnitudes while the minimum at secondary eclipse is 10.58 magnitudes. This means that at principal minimum the more massive component is in front of the less massive component, and this in turn means that the surface brightness of the more massive component is lower than the surface brightness of the less massive component: $J_1/J_2 < 1$. This result, though unusual, is not unprecedented among spectroscopic binaries in general. But in the case of AH Virginis, Chang concluded that the spectra are the same for both stars, namely K2, in the system of Morgan, Keenan, and Kellman as defined in their Atlas of Stellar Spectra. The amplitudes of the two velocity-curves in Figure 25 are very different, and the heavier component is approximately 2.4 times more massive than the lighter component. Yet both are visible in the spectrum and the true luminosities must be approximately the same. We have already discussed the resulting departure of the fainter component from the mass-luminosity relation and have explained it as the effect of radiation from a common gaseous envelope surrounding the entire system and carrying matter from the heavier component to the lighter, on one side, and from the lighter to the heavier, on the other.

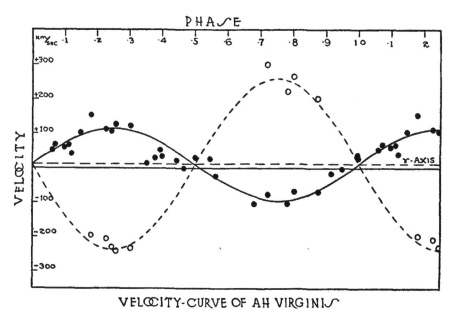

FIGURE 25.   Velocity-curve of AH Virginis by Y. C. Chang.

This situation is not unexpected, since we have observed a similar phenomenon in many early-type systems. For example, in Beta Lyrae there is a stream of gas passing from the more massive, luminous, B8-type component, whose spectrum we normally observe, to the less luminous F-type component, whose spectrum we do not see. This stream divides; part of it escapes into space in the form of a spiral ring, while another part passes around the cooler of the two components and finally returns to the original hot component. In the case of the W Ursae Majoris systems we have as yet no very clear understanding of the dynamical properties of the common envelope.

The W Ursae Majoris stars have been so named because W Ursae Majoris is the most typical representative of the group. It is a variable star with a period of almost exactly one-third of a day. There are two spectra visible, both of class F8. The apparent magnitude at maximum light is 8.0 and the two eclipses are about 0.7 magnitude and 0.6 magnitude in depth. The duration of the entire eclipse is 2 hours. There is no total phase, and it is believed that the eclipse is partial. The light-curve is rounded at maximum and the ellipticity of the components must be large. The period is known to vary, and there are also variations in the light-curve which resemble those that have been recently observed by O. J. Eggen in the case of 44 i Bootis (Figure 26). Similar variations were found by K. Walter in W Ursae Majoris. The spectrum was observed at the Mount Wilson Observatory in 1919 by W. S. Adams and A. H. Joy. They found that the eccentricity of the orbit is zero and that the semi-amplitudes of the two velocity-curves are 134 km/sec and 188 km/sec, respectively. The velocity of the system as a whole is -5 km/sec. The value of $a \sin i$ is 0.61×10^6 km for the more massive component, and 0.86×10^6 km for the less massive component. Finally, the two mass functions are $\mathfrak{M}_1 \sin^3 i = 0.67\mathfrak{M}_\odot$ and $\mathfrak{M}_2 \sin^3 i = 0.48\mathfrak{M}_\odot$. The star is also a visual binary and is listed in Aitken's double-star catalogue under the number 7494. The Mount Wilson observers gave no indication of the character of the spectrum, except that they noticed that the lines are diffuse and are suggestive of rapid axial rotation.

During the month of January 1949, a new series of spectrographic observations was obtained at the McDonald Observatory. The dispersion of the spectrograms was about 50 Å/mm and the general characteristics of the spectrum are very well shown (Plate XV). The duplicities of the absorption lines are pronounced and certain conclusions can immediately be made, even though the plates have not yet been measured. In a general way, the description by Adams and Joy is confirmed. The lines are strangely diffuse and lacking in contrast when they are double. They are broad and distinctly affected by axial rotation when they are single. But there are several other interesting features which had not previously been noticed. First of all, there is a conspicuous variation in the intensities of the two components. The violet component tends to be the stronger, irrespective of which star happens to be approaching. This phenomenon has been observed in every W Ursae Majoris system. The spectral type is later when the star is in eclipse (and the lines are single), irrespective of which component happens to be eclipsed. This is particularly noticeable in Ca I 4227. The latter is inconspicuous and weak when it is double, that is, during the two elongations; but it becomes very strong and dominates the spectrum when the lines are single, that is, during the two conjunctions. A similar effect has also been observed in the W Ursae Majoris-type system, SW Lacertae. Another feature may be of greater interest. When the lines are double, the sum of

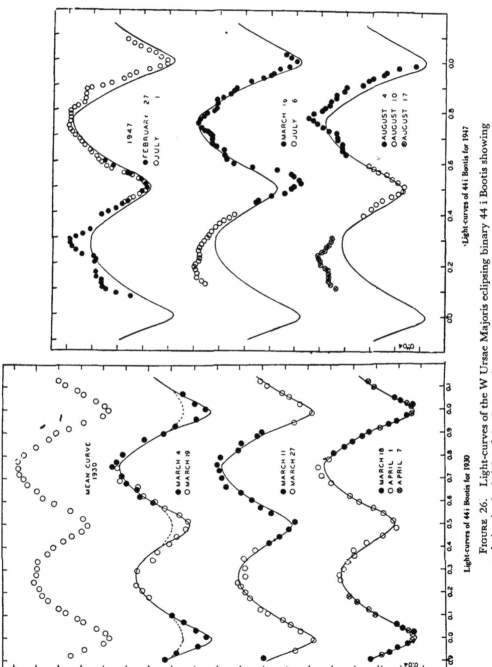

Light-curves of 44 i Bootis for 1947

Light-curves of 44 i Bootis for 1930

FIGURE 26. Light-curves of the W Ursae Majoris eclipsing binary 44 i Bootis showing variation in the height of the first maximum.

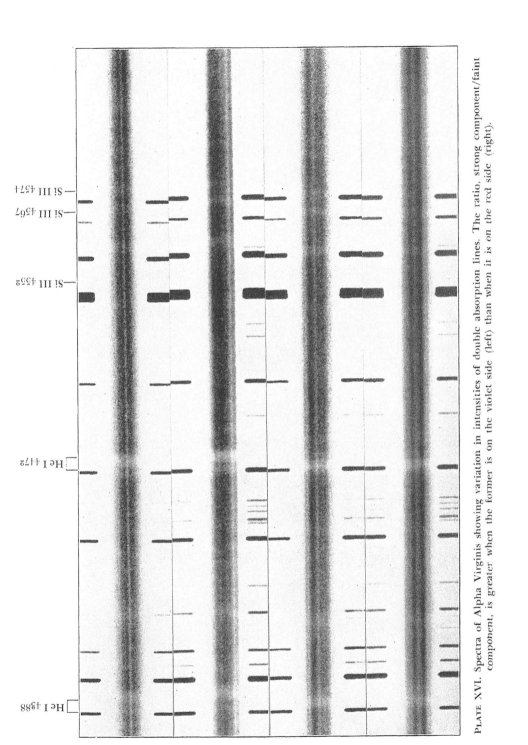

PLATE XVI. Spectra of Alpha Virginis showing variation in intensities of double absorption lines. The ratio, strong component/faint component, is greater when the former is on the violet side (left) than when it is on the red side (right).

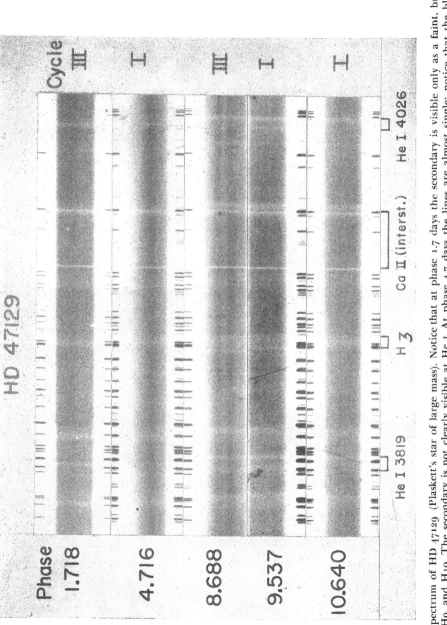

PLATE XVII. The spectrum of HD 47129 (Plaskett's star of large mass). Notice that at phase 1.7 days the secondary is visible only as a faint, broad feature on the red sides of Hξ, H9, and H10. The secondary is not clearly visible at He I. At phase 4.7 days the lines are almost single; notice that the blend Hξ + He I 3889 is weaker than H9 or H10. At phase 8.7 days of cycle III the lines are widely double, and the secondary is strong in He I; the latter is somewhat weaker and more diffuse in H; notice the strength of the violet component (secondary) for the blend Hξ + He I 3889. The velocity-curve predicts that at phase 9.5 days the separation of the components should be even greater than at phase 8.7 days. Yet, in cycle I, at 9.5 days the separation was much smaller than at phase 8.7 days of the same cycle. At the same time, the intensity of the secondary was exceptionally great. On the following night, phase 10.6 days, the secondary had become very weak.

their intensities does not add up to the intensity of the single line when the separation is zero. In other words, the lines are much deeper and stronger during the two eclipses than during the two elongations. During each elongation the whole spectrum gives the impression that there is continuous radiation, or possibly broadened line-emission, which fills in the double absorption lines and makes them lacking in contrast. It is not surprising that this effect has not been noticed before, because we know that in a spectroscopic binary with two components the continuous spectrum of one star overlies the line of the other and renders it less contrasty than it would be if the two lines were superposed. But in ordinary double-lined spectroscopic binaries which are not in contact, the weakening of the double-lined stage is exactly what we expect. For example, in the sharp-lined binary 47 Andromedae, the lines are stronger when they are single. They become half as intense when they are double; hence the two add up to give the intensity of the single line during each conjunction. This is true, irrespective of whether or not the stars have large rotational velocities. The binary 47 Andromedae does not have appreciable rotation. But in AO Cassiopeiae axial rotation is pronounced; yet the effect is of the same general character.

Apparently, some kind of radiation tends to fill in the double absorption lines of W Ursae Majoris at phases 0.25 and 0.75P. It is not unreasonable to suppose that this filling in results in some fashion from the common envelope which we have postulated in order to explain the strange departure of the weaker and the less massive component from the mass-luminosity relation (p. 26). This envelope produces absorption lines only when it is seen projected upon the nucleus of each star. Between them there will be gas at low pressure which will produce emission. One other point should be mentioned, namely, that although the lines are appreciably broadened, not only during the single-line stage but also during the double-line stage, this broadening is probably not as large as would be expected from the photometric elements of the system. If the two components were actually in contact, then, disregarding darkening at the limb, the line contours should touch. This is definitely not the case. The two spectroscopic components are well separated, and it is impossible to draw the contours in such a way as to obtain broad enough wings. It seems as though the two stars radiate as relatively small disks which are well separated from each other, despite the light-curve which proves that the two bodies are almost in contact.

In support of the conclusion that the sum of the intensities of the double lines during the elongations does not equal the intensity of the single lines during the conjunctions are the appearances of the very broad features in the spectrum, such as the lines of ionized calcium, H and K, and the broad CH band which is usually described as the G band. These features are so broad that, even with the large relative velocities of the two components in this system, their contours are not very greatly altered because of the motion of the two components with respect to each other. Yet they are much shallower, but not narrower, at the elongations than they are at the conjunctions.

The spectra show appreciable emission lines of Ca II in several of the W Ursae Majoris systems. It is well known that such emission lines are often observed in late-type binaries, and are quite conspicuous in some of them. They have been investigated by L. Gratton, W. A. Hiltner, and others, and have usually been found to come from a small region at the tidal bulges of one or both components. In W Ursae Majoris the Ca II emission lines are very broad and diffuse. They are best seen during both eclipses, when they are superposed over the very strong and broad, blended absorption lines. The emission lines move with respect to the underlying absorption line and can be interpreted by assuming that they belong to the more massive component of the binary system.

If we now turn to the spectroscopic observations of other W Ursae Majoris systems, we notice that there are a number of interesting regularities. Table XI shows the available data. In all well-observed cases there is a periodic change in the intensities of the two components. As we have already stated, the more massive component is eclipsed during the secondary minimum. The only exception to this rule is the system ER Orionis, in which the more massive component appears to be obscured during the primary eclipse. However, recent photometric observations make the two eclipses so nearly equal in depth that it is entirely possible that the primary and secondary eclipses will have to be interchanged when more accurate methods of measuring the light intensity have been applied to this star. Physically it is not quite easy to understand why this should be so, but observationally we have at least in one case an independent confirmation; in the star V 502 Ophiuchi, L. Gratton found that the spectrum of the heavier component is a little later than the spectrum of the less massive component (Figure 27).

We have already commented upon the mass ratios of the binaries. Among the ten systems which have been investigated to date, these mass ratios range from 1.2 in the case of SW Lacertae to 3.1 in the case of VW Cephei. Thus the dispersion in the mass ratios is quite large. A value of 1.2 would not be considered anomalous, but a value of 3.1 is extremely unusual. The mean is approximately 2. On the average, the masses of the two components, after applying corrections for $\sin^3 i$ to the mass functions, would amount to about one solar mass for the heavier component and one-half solar mass for the lighter component.

The two together would produce a star of 1.5 solar masses, provided that in the process of coalescing no mass is lost. If such a star remains on the main sequence, it would correspond to a region in the H-R diagram where the stars do not have rapid rotations.

The fifth column of Table XI gives the radial velocities of the centers of mass of the ten systems. They show the solar motion with respect to the local swimming hole—about 20 km/sec in the direction

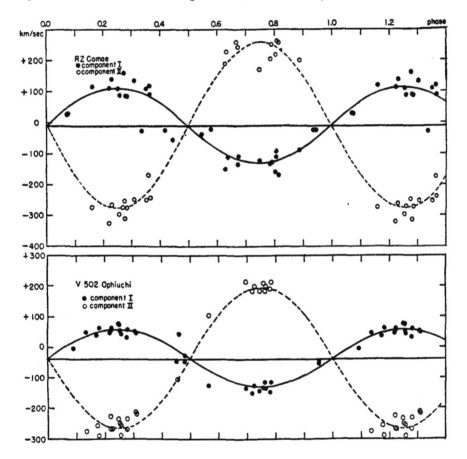

FIGURE 27. Velocity-curves of two systems of the W Ursae Majoris group.

of Hercules—and they also show individual motions, after corrections for the component of the solar motion have been applied, which are similar in dispersion to those of single stars of the same total mass. They are thus normal members of the H-R diagram, except that they are double. The next to the last column records whether the violet component is always the stronger in the spectrum, as in W Ursae Majoris. The observations are incomplete for three systems. In all the rest the effect is pronounced. Finally, the last column records the ratio of the surface brightness of the more massive (J_1) to that of the less massive component (J_2).

7. Variable Line-intensities in Spectroscopic Binaries

In the preceding section we discussed the remarkable tendency of the W Ursae Majoris systems to have strong violet absorption components and weak red components, irrespective of whether the more massive or the less massive star is approaching. The lines of at least one of the two stars are therefore variable in intensity. This phenomenon is related to an observation which was made many years ago at the Harvard Observatory. S. I. Bailey found, in 1896, that the relative intensities of the two components in the spectrum of Mu-one Scorpii undergo periodic variations. In 1901, Annie J. Cannon concluded that the difference in intensity between the two components is greater when the fainter lines have the longer wave length than it is when they have the shorter wave length. The entire collection of the Harvard objective-prism spectra of Mu-one Scorpii was discussed by Miss Maury in 1920. Her estimates indicate that, on the average, the intensity of the fainter component of several He ɪ lines is 1/3 that of the stronger component when the latter is approaching, and about 8/10 when it is receding. Although it is always possible to tell which component is the stronger and which the weaker, at certain times the weaker component is hardly visible, while at other times it is almost as strong as the stronger component.

Precisely the same thing was found at the Yerkes Observatory in the spectrum of Alpha Virginis. The fainter component in this system is always difficult to observe, but it is much stronger when it is on the violet side of the principal component, and it is faint and difficult to see when it is on the red side of the principal component (Plate XVI). This effect is also present in the spectroscopic binaries Sigma Aquilae, Beta Scorpii, AO Cassiopeiae and HD 47129 (Plate XVII). The evidence is so conclusive that no doubt can now be entertained as to its reality despite the fact that it is not easy to observe and that for many years astronomers were not all convinced of it. Petrie at Victoria confirmed this type of variation in one or two systems by means of accurate measurements. On the other hand, McDonald observations by C. T. Elvey and O. Struve did not confirm the early Harvard observations of Mu-one Scorpii, and D. M. Popper was unable to see any variation in the spectrum of V Puppis, which had also been announced by the early Harvard observers as showing this kind of change.

It is very interesting that this phenomenon exists in both groups of

TABLE XI

Systems of the Type of W Ursae Majoris*

No.	Star	Sp.	P days	γ km/sec	K_1 km/sec	K_2 km/sec	$\mathfrak{M}_1$	$\mathfrak{M}_2$	α	Int.	J_1/J_2
1	W UMa........	F8	0.334	−5	134	188	$0.9\mathfrak{M}_\odot$	$0.6\mathfrak{M}_\odot$	1.4	var.	<1
2	TX Cnc........	F8	.383	+36	112	217	1.5	0.8	1.9	"	<1
3	44 i Boo B.....	G2	.268	+3	116	232	1.0	0.5	2.0	?	<1
4	ER Ori........	G2	.423	+35	95	155	0.6	0.4	1.6	?	>1
5	YY Eri........	G5	.321	−20	130	200	1.0	0.7	1.5	var.	<1
6	VW Cep.......	K0	.278	−35	75	230	1.1	0.35	3.1	"	<1
7	AH Vir........	K0	.408	+10	105	250	1.5	0.7	2.4	"	<1
8	RZ Cam.......	K0	.339	−12	130	270	1.6	0.8	2.1	"	<1
9	V 502 Oph.....	G2 + F9	.453	−37	95	235	1.2	0.5	2.5	?	<1
10	SW Lac........	G3	.321	−22	172	202	1.2	1.0	1.2	var.	<1
	Mean				110	220	1.1	0.6	2		

* Reference: 1. Joy, *Ap. J. 49*, 189, 1919; Struve and Iwanowska (unpublished); 2. Popper, *Ap. J. 108*, 490, 1948; 3. Popper, *Ap. J. 97*, 407, 1943; 4. Struve, *P.A.S.P. 56*, 34, 1944; 5. Struve, *Ap. J. 106*, 92, 1947; 6. Popper, *Ap. J. 108*, 490, 1948; 7. Chang, *Ap. J. 107*, 96, 1948; 8. Struve and Gratton, *Ap. J. 108*, 502, 1948; 9. Struve and Gratton, *Ap. J. 108*, 502, 1948; 10. Struve, *Ap. J.* (in press).

close binaries, the W Ursae Majoris systems and the early-type contact binaries. Otherwise, the spectra are entirely different. Undoubtedly we are concerned with a property of all close systems, and its explanation forms an intriguing problem.

But we must first consider the rather strange lack of consistent results in the case of Mu-one Scorpii and V Puppis. A partial explanation, or analogy, may perhaps be found by referring to recent photoelectric determinations of the light-curves of close eclipsing binaries. Eggen's light-curves of i Bootis (Figure 26) show remarkable changes in the maximum brightness of the first hump in the curve. It is not yet certain how these variations take place. He had tried to match the higher of the two maxima. By doing so he was able to fit fairly closely the principal minimum and, to a lesser degree, the secondary minimum. Having thus obtained a more or less satisfactory fit for the higher of the two maxima and for the two minima, he noticed that the secondary maximum, at his phases 0.2 to 0.3, gave different heights for different series of observations. For example, in the curves in Figure 26 which were obtained from observations by A. E. Whitford at the Mount Wilson Observatory in 1947, the points for various dates between February 27 and August 17 give highly discordant results when compared with the curve which J. Stebbins had obtained in 1930. In the earlier group of observations the two maxima were different in height; in the more recent group they were very nearly the same. Although it is not evident what physical connection there may be between the total brightness of the star at each maximum and the relative strengths of the two absorption components in the spectrum, in all probability there exists a variable factor in the atmosphere of i Bootis which simultaneously causes the star to appear different in brightness and different in the ratio of the two absorption components.

By a strange coincidence, the same issue of the *Astrophysical Journal* (July 1948) which contained Eggen's article on i Bootis, contained also an article by F. B. Wood on the light-curve of AO Cassiopeiae— one of those O-type spectroscopic binaries in which the variation of the line-intensities is prominent. Wood was able to combine his own photoelectric observations with the light-curves of several earlier workers by means of the expression

$$l = 0.9284 - 0.0578 \cos 2\theta + x \sin \theta + y \sin 2\theta,$$

where l is the apparent luminosity of the combined light of both stars outside of eclipse. The term in $\cos 2\theta$ represents the usual effect of the ellipticity of the components. If the angle θ is the longitude of the star counted from primary mid-eclipse, this term produces a double wave, with minima when the apparent disks of the stars are small and with maxima during the two elongations, when we see the broad sides of the components. The term in sin 2θ produces a phase shift of the ellipticity effect and has no physical explanation. The term in sin θ produces a difference in the heights of the two maxima, and it is this term which resembles Eggen's results for i Bootis. There is no physical explanation for it. The coefficients x and y are variable in such a way that sometimes $x = 0$, while at other times it has a negative value. In Wood's own observations $x = -0.0118$, making the first maximum lower than the second. But in A. Bennett's observations, in 1933, the curve was symmetrical, while in W. A. Hiltner's observations in 1948–1949, the constant x was very small.

If, as we surmise, the sin θ term in the light-curves of close eclipsing variables is related to the variation of the spectral lines, then perhaps the variability of the coefficient x gives a clue to the puzzling lack of consistence among the spectrographic results. Miss Maury had remarked, long ago, that the variability of the lines in Mu-one Scorpii was not always noticeable, and recent work also suggests that the range of the line-variability undergoes changes. Measures of this phenomenon have, thus far, been made only in very few systems. In the case of Alpha Virginis (Plate XVI), where it is not particularly striking, an effort has been made to obtain quantitative measures of the variation, not only of the faint component, but also of the strong component. In the case of the former, the ratio between the line-intensity near its velocity minimum to the line-intensity near its velocity maximum is 1.7. For the strong component the same ratio is 1.4. Three possible explanations suggest themselves:

(*A*) The phenomenon resembles that postulated many years ago by J. C. Duncan for the explanation of the light-curves of Cepheid variables. This hypothesis has been abandoned, in so far as the Cepheids are concerned. The two components are supposed to have identical periods of rotation and orbital revolution. Their motion in a resisting medium would heat the advancing side of each, thus causing a variation in the intensities of the two continuous spectra.

According to this hypothesis, the variation is entirely due to changes in the continuous spectra and not to the lines themselves. The amount of absorbing matter is considered equal in all four spectra. This presupposes that the spectral types of the two components are the same. Let the total absorptions of the lines in terms of their continuous spectra be $A'_1 = A'_2 = A''_1 = A''_2 = A$, and let the intensities of the continuous spectra of the two sides of component one be C'_2 and C'_1, while the intensities of the continuous spectra of component two are designated C''_1 and C''_2, for the fainter and the brighter side, respectively. The observed continuous spectrum is $C'_1 + C''_1$ when the velocity of the primary is negative and $C'_2 + C''_2$ when the velocity of the primary is positive. The observed intensities of the four lines are

$$\frac{AC'_1}{C'_1 + C''_1} = a; \quad \frac{AC''_1}{C'_1 + C''_1} = b; \quad \frac{AC'_2}{C'_2 + C''_2} = d; \quad \frac{AC''_2}{C'_2 + C''_2} = e.$$

The measurements give

$$\frac{a}{d} = 1.4; \quad \frac{e}{b} = 1.7.$$

Furthermore, we also find from the measurements:

$$\frac{a}{b} = 8.8.$$

It is, of course, obvious that the observations can give only the two ratios C'_1/C''_1 and C'_2/C''_2. We cannot derive C'_1/C'_2. In other words, the observed intensities depend upon the ratios of intensity of the bright side of the primary to the faint side of the secondary, and of the faint side of primary to the bright side of the secondary; but they tell us nothing concerning the quantity $m = C'_1/C'_2$. If m is known from other data, then we also know C''_1/C''_2.

It is seen that

$$\frac{C'_1}{C''_1} = \frac{a}{b} = 8.8.$$

Forming

$$\frac{a}{d} = 1.4 = \frac{\left(1 + \dfrac{C''_2}{C'_2}\right)}{1.1} \quad \text{and} \quad \frac{e}{b} = 1.7 = \frac{9.8}{\left(\dfrac{C'_2}{C''_2} + 1\right)},$$

we find two values of C''_2/C'_2:

$$\frac{C'_2}{C''_2} = 1.8; \quad \frac{C'_2}{C''_2} = 4.8,$$

which are inconsistent with one another. We conclude that the observations are not consistent with the hypothesis $A'_1 = A'_2$; $A''_1 = A''_2$. This may, however, be due to the lack of precision in our observations.

We shall tentatively adopt $C'_2/C''_2 = 3.5$. Supposing, next, that the stronger component does not vary at all ($m = 1$), we find $C''_2/C'_1 = 2.5$. The bright side of the companion is 2.5 times more intense than the faint side.

The spectral class of each of the two components is B2, corresponding roughly to $T'' = 15,000°$K. Computing from Planck's law the temperature corresponding to an increase in $\mathcal{J}(\lambda) = c_1\lambda^{-5}(e^{c_2/\lambda T}-1)^{-1}$ by a factor of 2.5, we find for $\lambda = 5000$ Å,

$$T' = 25,000°K.$$

Such a large difference between T' and T'' should produce a marked difference in the spectral types of the two sides of the companion, which is contrary to observation.

Another objection to hypothesis (A) is the mechanical difficulty of accounting for so large a difference in temperature. The energy required is so enormous that the density of the resisting medium would have to be much greater than could be reconciled with the fact that the period has been constant to within at least 0.00001 day (or 1 second) in an interval of forty-four years.

(B) It might be suggested that the variation of the intensities is caused by the reflection effect in a close binary. Eddington has given a formula for computing the amount of light reflected from the companion, when we know C'/C'' and $\phi = R_2/a = $ radius of companion/ distance of centers. For several eclipsing variables the brighter side of the companion is twice as intense as the fainter side. For Alpha Virginis, R_2/a is not accurately known, but assuming a reasonable value for this ratio, it seems probable that the reflected light would make an important contribution to the total brightness of the companion.

If the two components are spherical, or if they are ellipsoids having their major axes in the line joining their centers, the reflection effect at both elongations is the same and the fainter component should be strongest at opposition. The observations prove that the effect at the two elongations is not the same.

If there is no equality in the periods of rotation and revolution, tidal lag might cause the major axes to be displaced from the line joining the centers. In that case the reflection effect would not be the same in the two elongations. However, maximum and minimum line-intensity should occur somewhere between elongation and eclipse, and not at elongation as observed. Moreover, a large amount of tidal lag is improbable, according to the theoretical work of Z. Kopal.

Furthermore, the ellipticity of the components of close binaries is usually small, and while the reflection effect in the companion may be observable in opposition, it will produce but little difference in the two elongations. The ellipticity of Alpha Virginis has not been determined, but J. Stebbins found a probable variation in light amounting to about 0.10 magnitude. Judging from the somewhat similar case of the variable Pi-five Orionis, for which Stebbins finds an elongation of only a little more than five per cent, we can conclude that the ellipticity of the stronger component in Alpha Virginis is insufficient to cause the observed variation in the line-intensities. There remains, however, the possibility discussed by K. Walter, that the ellipticity of the fainter component is different from that of the stronger.

(C) The third hypothesis attributes the changes to an actual increase in the amount of absorbing gas on the advancing side of each component. There is no theoretical explanation for this increased thickness, but, accepting it as a working hypothesis, we estimate that

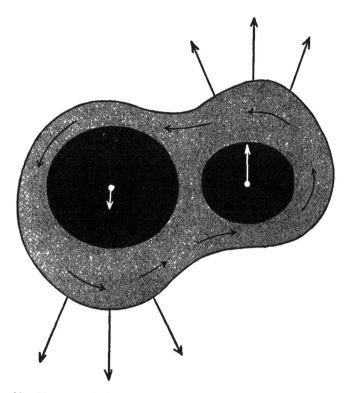

FIGURE 28.   Unsymmetrical common envelopes in close binary systems.

the amount of material on the advancing side of the reversing layer of each component is roughly twice that on the receding side. Perhaps we must think of these stars as being surrounded by an unsymmetrical common envelope of very tenuous gas, having a greater thickness in front of the leading hemisphere of each star. This has already been rendered probable by the departure from the mass-luminosity relation of the W Ursae Majoris systems, but it is not yet clear how this envelope has originated. G. P. Kuiper has shown in his study of contact binaries that when an envelope exists around two close stars of unequal mass, a current will be set up in it which moves matter from the more massive star to the less massive star, and this current may, as it does in the case of Beta Lyrae, split into two branches: the one escapes permanently from the system and disappears as a spiral of diffuse nebulous matter, while the other circumscribes the less massive component and returns to the source of its origin. But we ask: what produces the common envelope in the first place? Perhaps we are concerned with prominence activity on a very vast scale. We shall see that in UX Monocerotis prominences are produced predominantly on the advancing hemispheres of the two stars, and that we may have as the result an unsymmetrical envelope in which the spectral lines produced by the advancing side will be relatively stronger than the spectral lines produced by the receding hemisphere of each component (Figure 28).

8. Gaseous Rings in Close Binary Systems

The problem of the unsymmetrical envelopes in spectroscopic binaries is closely related to that of luminous streams of hydrogen and other gases which were discovered by A. H. Joy in the system of RW Tauri and have since been found in a large number of other close pairs.

In 1934 a significant observation was made by the late Arthur B. Wyse of the Lick Observatory. In a footnote to his well-known paper "A Study of the Spectra of Eclipsing Binaries," he noted that on an exposure made about four minutes after mid-eclipse of the variable RW Tauri "the plate showed an A-type spectrum, on which were superimposed the bright lines $H\beta$, $H\gamma$, $H\delta$, and probably λ 4481 of Mg II. The emission lines were displaced to the violet edges of the absorption lines." On a later plate taken at the same phase and under the same conditions "no bright lines were visible, although the absorption spectrum was present with the same intensity as before." The variable was again observed in 1940 and 1941 by A. H. Joy with the 100-inch telescope of the Mount Wilson Observatory. The bright lines were found to consist of two components: "at second contact the red emission component appears alone, and at third contact the violet component, while at mid-eclipse no emission is seen." Joy suggested that these "emission lines originate in an extended gaseous ring surrounding the equatorial region of the primary B9 star." The larger K0 star first occults one-half of the rapidly rotating edge-on ring—namely, that half which approaches the observer. For a short interval of time the K0 star eclipses the entire ring and no bright lines are visible. This, of course, happens when the B9 star is also totally eclipsed. Finally, the approaching side of the ring is uncovered, and the receding side alone is occulted. Observations of this type are difficult, because the star is faint and the entire eclipse lasts only 6.5 hours, while totality lasts 84 minutes. The period of RW Tauri is 2.77 days.

During the past few years an effort has been made at the McDonald Observatory to study the spectroscopic characteristics and to determine the radial-velocity curves of some 100 faint eclipsing variables. Approximately two-thirds of these stars have total or nearly total eclipses. Of these, again about two-thirds have been studied with sufficient care during the eclipse. Among these stars—roughly 30 in number—about 15 have been found to possess bright lines of hydrogen and sometimes of other elements, which vary in intensity in precisely the same manner as do the bright lines of RW Tauri. The observations are difficult: most of the variables investigated have photographic magnitudes between 10.5 and 13.5 at principal minimum; the eclipses last only a few hours, and the changes in the bright lines caused by the occultation of the gaseous rings by the large late-type components are often very rapid; the predicted times of mid-eclipse are occasionally in error by several hours so that an accurate determination of the epoch of mid-eclipse must sometimes precede the spectrographic observations.

Despite these difficulties we find that about one binary in three having deep total eclipses possesses easily observable features associated with the existence of gaseous rings in these systems.

We turn to the description of the principal characteristics of the gaseous rings as they have been revealed by the spectrographic observations. In all, 15 stars are now known to possess this feature, but only 12 have been fully investigated. Their periods range from 2.77 days for RW Tauri (investigated by Joy) to 261.9 days for RZ Ophiuchi (investigated by W. A. Hiltner from spectrograms secured at the McDonald Observatory). The frequency-curve of all eclipsing stars has a steep maximum for periods between one and two days. The rings clearly prefer longer periods. This may to some extent be caused by the difficulty of observing the phenomenon when the eclipse is of very short duration, but it is fairly obvious that this effect of selection does not nearly account for the observed preference of the longer periods.

Next, we find that there is a striking relation between period and spectral type. The shortest period is associated with class B9 and the longest with class F5. The correlation is nearly perfect: for example, RW Persei and RS Cephei, with periods of 13 and 12 days, respectively, both have spectral types A5. An interesting case is that of RX Cassiopeiae, which has usually been classified as G3. It turns out that the G3 component, although it is the more luminous in photographic light, is not the one of earlier type. There is another component, of about type A3, and it is this earlier component which is associated with the ring. In RX Cassiopeiae, as in all the other systems, the star of earlier spectral type is the one eclipsed when the emission lines show the characteristic variations in intensity. The rings are associated with these components, at least in the sense that they produce bright lines in the vicinity of these stars.

The rings show a marked velocity of rotation, and in all cases the direction of motion is the same as that of the binary system. But the velocities are much larger: for RW Tauri the velocities of the ring are $+350$ km/sec. They are correlated with the period, and lead to a relation of the form $V^3_{em} \propto P^{-1}$. If the masses of the systems were all the same, this would be equivalent to Kepler's third law, but it must be remembered that the masses are not the same but that they probably increase with period because those of longest periods possess marked giant or supergiant characteristics. Moreover, the P is not the period of the rings but that of the binary system. Probably the similarity to Kepler's law is somewhat accidental. But what we do have is this: for the longer periods, the space available for the rings is larger than for the shorter periods.

The eclipses of the rings tend to show that the latter are roughly of the same order of size as the cooler, larger, eclipsing components of the binaries. The hotter components are two or three times smaller. This does not mean that the rings are all about the same size. The systems of longer period show giant or supergiant characteristics in the spectra of their hotter components, and it is probable that the cooler components are also systematically larger in such stars as RZ Ophiuchi than in RW Tauri. In addition, there is a remarkable tendency for the rings to be totally eclipsed in the short-period systems, while the eclipses are the equivalent of annular in the long-period systems. For example, in RW Persei, RX Cassiopeiae, and SX Cassiopeiae the bright components become approximately equal and relatively weak at mid-eclipse (that is, they are weak if we free their intensities from the effect of prolonged exposure at the time of eclipse), but they never completely disappear as they do in RW Tauri and in some of the other short-period systems.

The total luminosities of the rings are also greatest in the systems having the longest periods. This is undoubtedly due to the fact that in the larger systems the volume occupied by the rings is relatively larger than the volume occupied by the component stars. In consequence of this, we observe stronger emission lines in the long-period systems, and in some the lines are visible even in full light. On the other hand, all short-period systems with observable rings are characterized by very deep total eclipses. The ranges are 3.7 magnitudes for RW Tauri, 2.5 for SW Cygni, 2.2 for W Delphini, 2.7 for AQ Pegasi, 2.5 for TT Hydrae, 2.6 for VW Cygni, 2.0 for RY Geminorum, 1.7 for RS Cephei, 2.2 for RW Persei, 0.7 for RX Cassiopeiae, 1.0 for SX Cassiopeiae, and 0.8 for RZ Ophiuchi. A short-period system with a shallow eclipse would rarely, if ever, show the bright lines. On the other hand, in moderately long-period systems the emission lines are comparable in photographic density to the neighboring continuous spectrum of the early-type component.

The bright-line spectra of the rings are quite similar to one another. The H lines are always the strongest. If they are very strong, we also observe Fe ii, Ca ii, and Mg ii 4481 in emission. There are no forbidden lines. We may say that the state of ionization of the rings is approximately the same as that of the exciting early-type stars. But the H lines show no broadening of the kind produced by Stark effect, and the pressures must be very low. The total monochromatic luminosities of the rings are quite appreciable. In several binaries of long period, the intensities of the $H\beta$ components are comparable with those contained within about one or two Ångstrom units of the neighboring continuous spectrum of the blended light of the two stars. This is only a rough estimate, but it serves to show that the integrated light of such a ring within the range of sensitivity of an ordinary photographic plate is of the order of perhaps 10^{-3} of the combined light of the two stars. If we recall that the size of the ring is not very different from that of the solar corona, we realize that in luminosity, relative to its total star light, the ring is about a thousand times brighter than the corona is relative to the sun. It is easy to determine the number of hydrogen atoms in the ring, but for the present purpose it is sufficient to point out that the strongest H lines of the rings are comparable in intensity to the bright H lines in Be stars. But, since the former are measured relative to the continuous spectra of A-type stars while the latter are measured relative to the continuous spectra of early B-type stars, the total numbers of H atoms in the higher quantum states must be smaller in the rings than in the Be stars.

The motions within the rings must be very complicated. We have some slight observational evidence concerning these motions in the cases of SX Cassiopeiae and RX Cassiopeiae, because in these two stars we have observed the bright lines throughout their entire cycles, not only at the time of eclipse. It turns out that the emission lines vary in character with the phase, and their central absorption cores— undoubtedly also produced within the rings—disappear at one or both elongations. The inference to be drawn from these observations is that the motions are not simply circular, around the hotter components, but may well be as complicated as the motions of the gaseous spiral in Beta Lyrae. Even the hypothesis of a ring may turn out to be only a crude approximation.

An interesting problem is presented by those stars which do not possess observable rings. We notice that not one of the systems with emission lines has a spectral class earlier than B9 or later than F5. It is probable that this is an effect of ionization; the hotter stars ionize the material of the ring to such an extent that emission lines of H are not observed. The cooler stars fail to excite the atoms of H. But there are systems of class A with deep eclipses—such as U Cephei, SX Hydrae, UU Ophiuchi—which show no emission lines. We notice that the periods of these stars are all near the shorter limit of 2.77 days set by RW Tauri. Thus there are real differences in the luminosities of the rings, a fact which is apparent also when we compare the relatively weak emission lines of W Delphini with the much stronger ones of AQ Pegasi. Joy has shown that the rings of RW Tauri are not stable formations. The total intensities of the emission lines vary in an irregular manner, as does also the ratio: violet component to red component. Ring-formation is a very common process in binary systems, perhaps even a universal one, and it is only because of our instrumental and other limitations that we are restricted in our observations to the discovery of the most luminous rings.

The interpretation of the double bright lines in eclipsing binaries, which we have discussed in the preceding sections, is strikingly similar to that advanced many years ago in connection with the Be stars. In the latter we have no such direct observational evidence of the existence of a ring as we have in the phenomenon of occultation of the opposite halves of the rings in eclipsing binaries. But the evidence in favor of ring-like structures surrounding normal B-type stars is nevertheless very strong.

In view of this similarity between the Be stars and our group of eclipsing binaries, it is interesting to compare the two phenomena in detail. In profile there appears to be no significant difference between the double emission lines of the binaries and those of the Be stars. For obvious reasons we have not as yet discovered binary systems with single narrow emission lines, but we know many Be stars with single emission lines. It is probable that this difference is due entirely to the absence of eclipses in the corresponding group of binaries. There must be stars like SX Cassiopeiae or RZ Ophiuchi whose inclinations are near 0° and whose emission lines are strong enough to show without eclipses, but such stars would arouse little interest. Perhaps some of P. W. Merrill's Ae and Fe stars are such objects.

The spectra of the two groups differ in several important respects:

(A) The stellar spectra are of early type B in one group, and of types B9 to F5, with the middle A's predominating, in the second.

Yet the emission-line spectra show approximately the same state of ionization.

(B) The broad double emission lines in Be stars are invariably associated with rapid axial rotation of the stars. In the binaries there is no such connection; SX Cassiopeiae and RZ Ophiuchi are distinguished for their sharp and narrow absorption lines. In the other binaries the absorption lines are not quite so narrow, but they certainly do not show any very large stellar rotations.

(C) It is perhaps characteristic that in those binaries in which the emission spectra are conspicuous, Mg II 4481 is usually reported as being present, along with Fe II and Ca II. In the Be stars, the Mg II emission line is either absent or exceptionally weak. Since this line is very sensitive to the dilution of the exciting radiation, in emission as well as in absorption, this may possibly indicate a greater degree of dilution in the Be stars than in the binaries. But in the binaries, as well as in the Be stars, the bright components of Mg II 4481 often look abnormally broad and diffuse. In the case of the Be stars this phenomenon suggested that the Mg II lines are produced at a lower level than the lines of H and Fe II, where the velocity of rotation is greater than at the higher levels, this being a consequence of the law of conservation of angular momentum. It is reasonable to adopt the same explanation in the case of the binaries.

(D) It should be emphasized that the ordinary Be stars are not often known to be spectroscopic binaries. In fact, it seems fairly certain that the great majority of them are ordinary single rapidly rotating B stars.

We encounter an interesting complication when we attempt to explain why the ionization of the two kinds of rings is approximately the same. Consider hydrogen with its ionization potential of 13.5 volts. In the absence of a better theory, we shall follow the usual procedure and assume that the ionization is produced by diluted black-body radiation from the hotter stars, so that we may write

$$\log \frac{n^+}{n} = -\chi \frac{5040}{T} + 1.5 \log T + \log W - \log n_e + C.$$

The observations indicate that n^+/n is the same for both groups o objects. Hence, if we introduce

$$\chi = 13.5; \quad T_{Be} = 20{,}000°; \quad T_{Binaries} = 10{,}000°,$$

we find:

$$\log \left(\frac{W}{n_e}\right)_{Binaries} - \log \left(\frac{W}{n_e}\right)_{Be} = 3.6.$$

The dilution factor, W, has been estimated for the Be stars to be of the order of perhaps 10^{-2}. That for the binaries is, if anything, nearer to one, if we can rely on the evidence of the Mg II line. Let us suppose that it is of the order of 10^{-1}. Then our result would mean that the electron density of the rings differs by a large factor from that of the rings in Be stars; the electron pressures in the rings of binaries must be smaller than in those of Be stars.

The gaseous rings produce absorption phenomena as well as emission lines. We have already commented upon the central absorption cores of the H lines and their variations in such stars as SX Cassiopeiae, RX Cassiopeiae, and RW Persei. There are other even more important effects. The strangest is the influence of the absorption lines of the rings upon the velocity-curves of the binaries. This is strikingly shown in the case of SX Cassiopeiae, which has a very unsymmetrical velocity-curve suggesting a large eccentricity and a value of ω close to $0°$. Yet we know from the light-curve that the eccentricity is probably very small. All other systems having rings, which have thus far been investigated, show similar velocity-curves, and most of them are incompatible with the light-curves.

It is remarkable that, even in some systems which do not show emission lines, the velocity-curves are distorted, and there is reason to believe that gaseous streams, or rings, are responsible for these distortions (p. 229). Among the most outstanding examples of this sort are U Cephei and AU Monocerotis.

The great frequency and luminosity of the rings in systems of fairly long period shows that they are not limited to pairs in contact, like Beta Lyrae. Since the components of these systems have slow rotational velocities, the rings are not now being formed through rotational breakup, as is probably the case in the Be stars. But Joy has shown that the rings in RW Tauri are not always the same; hence they are probably unstable formations being constantly replenished through prominence action from the stars, and being dissipated into space (as in Beta Lyrae) through turbulent motions.

Joy's original explanation (Figure 29) envisages a symmetrical ring around the hotter star. But it is possible that the observations can be explained equally well by postulating an envelope, as in Figure 28, which revolves very rapidly in the same direction as the orbital motion and in which the hotter component ionizes the hydrogen in its vicinity, thereby causing recombination, and hence emission lines only from those parts which are closest to its hot surface.

9. Typical Spectroscopic Binaries

During the past ten years a vast amount of material has been accumulated concerning the motions of the gases in the common envelopes of close binaries. When the envelopes produce emission lines, as they do in the binaries we have just discussed, the problem is fairly easy because the normal reversing layers of the stellar components have only absorption lines, and blending does not complicate our measurements. But the absorption phenomena in the envelopes are more difficult to disentangle. As a rule, the state of ionization in a common envelope (or shell) differs only slightly from that of the reversing layer of the brighter star. It is easy to explain this: we believe that the gases in the envelope originate in the form of promi-

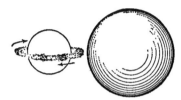

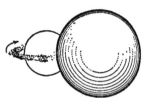

a. Large, faint *K*0-type star about to eclipse smaller, more brilliant *B*9 star with luminous ring of hydrogen gas surrounding it. Spectroscopic analysis of light shows only presence of bright *B*9 star.

b. The *B*9 star, although partially eclipsed, still so far outshines the *K*0 star that spectroscope reveals only usual *B*9 spectral characteristics.

c. At moment when *B*9 star is completely blocked out, brilliant bright lines of hydrogen gas flash out. These bright lines show Doppler shift to *red* side of normal position, indicating velocity *away from* observer of 210 miles per second.

d. The *B*9 star and its ring are both totally eclipsed for thirty minutes. During this phase spectroscope shows only very faint solarlike spectrum of the *K*0 star.

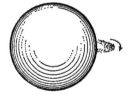

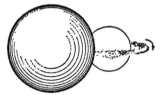

e. At third contact, ring on other side is revealed and spectroscope now shows bright hydrogen lines again, but this time shifted to the *violet* by an amount indicating velocity of 210 miles per second *toward* the observer.

f. Spectrum is now same as that in *a.*, a typical hot *B*9 star only. Entire process repeated in 2 days 18 hours.

FIGURE 29. Joy's interpretation of the emission lines in RW Tauri. (From *Astronomy*, by W. T. Skilling and R. S. Richardson, Henry Holt and Co., 1947.)

nences from the stars. If, as a first approximation, the velocity of each prominence, v, is constant, then the increase in volume which is filled when such a prominence spreads its atoms throughout the available volume after an interval Δt is equal to $(r/R)^2$, where $r = (v\Delta t + R)$. Hence, the density of the gas is reduced by the same factor, and the ionization, in thermodynamic equilibrium, would be increased in the same ratio. But the radiation which is responsible for the ionization is also reduced in density by precisely the same dilution factor, $(r/R)^2$. Thus, if the reversing layer blends into a vast field of prominences, with the gases effectively distributed as in a uniformly expanding shell, the ionization will be the same throughout.

The absorption lines in the envelope and in the reversing layer tend to produce blends, and it is often difficult to distinguish between them. This complication has retarded our progress, and has led to the construction of several working hypotheses and models which, at first sight, are quite different from one another. But they have a number of features in common, so we can now construct a composite model which combines these common features and disregards those which are not relevant. The best way to proceed is to outline several of the individual models and later to combine them into our working model.

Historically, the first model of a contact binary with a common envelope was Beta Lyrae. Although it has been known as a variable star since 1784 and although its spectrum has been investigated more minutely than that of any other star, the physical nature of this remarkable and unique object has not yet been fully explained. The spectrum consists essentially of three distinct sources: (1) A normal giant-type absorption spectrum of class B9, which shows a regular variation in velocity with a range of 367 km/sec and with a period of approximately 13 days. This source is identified with that component of the binary system which is obscured at primary eclipse. (2) An abnormal absorption spectrum resembling spectral type B2 or B5, varying in intensity, but showing little or no variation in velocity during the entire cycle of the binary. (3) An emission spectrum of variable intensity corresponding to the B5 absorption lines and consisting of broad emission bands which have been suspected by some, but not by all, observers to vary in a peculiar manner in wave length and in radial velocity. It was generally believed, until a few years ago, that the B2 to B5 absorption lines represent the spectrum of the second component of the binary, and Miss A. C. Maury at the Harvard Observatory derived for this component a very large mass, because its variation in radial velocity was exceedingly small. However, later investigations have shown that the B5 absorption spectrum does not represent the spectrum of a star, but that it originates in an expanding nebulous mass of gas which is located somewhere between the bright B9 component and the observer.

Thus we observe in the spectrum the light of only a single star. We do not have any direct spectroscopic information concerning the fainter component in the system. It is strange that for a great many years the photometric observers were unable to tell which of the two components was the brighter and which the larger, but now we know that the B9 star is larger and more massive than the other, invisible, component, whose color is that of a star of spectral class F.

In order to understand the physical phenomena which play a role in the spectrum of Beta Lyrae, it is important to consider the B5 spectrum which originates in the nebulous mass. This spectrum undergoes large variations in intensity. When it is strong it shows a weak line of Mg II 4481 and probably very weak lines of Si II 4128 and 4131. These lines are very sensitive to the effect of dilution of radiation. When the radiation is that of thermodynamic equilibrium, these lines have normal intensities. But when the source is at a great distance from the exciting atoms, so that the radiation does not come uniformly from all parts of the sphere around them, the populations of the lower levels from which these lines arise are greatly reduced. We have already seen that the reason for this is found in the cyclic processes which lead to the various populations of the different levels (p. 141). In thermodynamic equilibrium these populations are determined by the Boltzmann relation, but under conditions which depart from thermodynamic equilibrium one has to consider the precise mechanism which gives rise to the population of any given atomic state. In the case of a gas at very low pressure which is being illuminated by the strong radiation from a neighboring star, the process which plays the predominant part is the excitation by radiation. The lower levels of the three lines we have mentioned are not metastable. They connect with the ground level of each atom by means of one or more strong transitions. As a result, the populations of these levels are greatly reduced if the radiation is not that of thermodynamic equilibrium. If a spectral line originates from a metastable lower level, then that level has no permitted transitions downward and it can be shown on the basis of the theory of cycles that its population remains very nearly the same as that which is found in thermodynamic equilibrium, even though the density of the available radiation is much less than that which would correspond to thermodynamic equilibrium. Now, in the case of the B5 component of Beta Lyrae, we have very strong lines of H and of He I. In any normal stellar spectrum having lines of helium similar to those of the B5 component in Beta Lyrae, we always expect to see a strong line of Mg II at λ 4481 and we also expect to see strong lines of Si II. In the B5 spectrum of Beta Lyrae these lines are quite weak, perhaps 10 or 100 times weaker than they would be in thermodynamic equilibrium. This leads to a determination of the dilution factor, $W = 0.05$. Now, the dilution factor measures the solid angle under which the surface of the star is seen from the point in the nebula which we are considering, measured in terms of the complete sphere. This angle is approximately given by the expression $R^2/4r^2$, where R is the radius of the star in linear units and r is the distance from the center of the star to the point in the nebula at which the absorption takes place. Hence, for a dilution factor of the order of 0.05, we simply substitute this quantity in place of $R^2/4r^2$ and solve for the ratio of r/R. This turns out to be approximately 3. We are thus concerned with a nebulous layer which is quite high above the photosphere of the B9 star.

We have as yet no definite means for measuring the thickness of

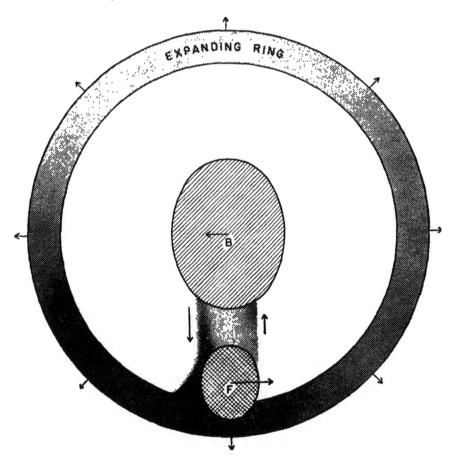

FIGURE 30. Working model of Beta Lyrae.

the layer which gives rise to the B5 absorption lines of H, He I, and a few other elements. But the layer cannot be very thick; otherwise we would not have been able to obtain a unique determination for the dilution factor. Figure 30 shows the outcome of the investigation. We see the two cross-sections representing, in the center, the B9 star, and below it the smaller and less massive F star. The arrows from the center of each star represent approximately the velocities in the circular orbit of each component. The more massive B9 star has the smaller velocity and, hence, the smaller arrow. The ring around the system represents approximately the location of the B5 layer. The shading of the layer shows approximately the strength of the absorption lines produced by it as seen from different directions. When we observe at principal eclipse, the direction of our vision is from the bottom of the page toward the top. As the eclipse progresses we continue observing the F-type star projected in front of the B9 star until, near the end of the eclipse, at phase approximately 0.1 P, no part of the B9 star remains obscured by the F star. It is approximately at this phase that we observe very strong absorption lines belonging to the B5 spectrum. We must be looking through a dense mass of gas, and this is shown by the blackness of the layer corresponding to the B5 component. In the diagram the B5 layer has little arrows pointing outward. This is intended to show that the radial velocities, as measured from the B5 lines, give a systematic velocity of approach, no matter from which direction we are observing the system. We actually observe as an absorption line only that part of the layer which at any given phase is seen projected upon the disk of the B9 star. Thus, if we observe at phase 0.25 P, we see the two stars side by side. This phase is one of elongation of the two components. The B9 star is then approaching us, and the F-type star is receding. At this same time we observe a certain portion of the B5 line in front of the B9 star, and it is this layer which shows a velocity of approach toward us and of recession from the B9 star. The layer is expanding with a velocity which averages about 75 km/sec.

Just before principal mid-eclipse the spectrum shows additional absorption lines which are strongly displaced toward the red side of the spectrum and which are not seen at any other part of the cycle. Immediately after mid-eclipse we see another set of absorption lines with a strong displacement toward the violet. These lines are not visible at any other stage of the cycle. They exist for only a very short time and they change rapidly from one to the other, just as though we were concerned with a phenomenon which is rapidly altered because of the obscuration produced by the F-type star near mid-eclipse. These particular highly displaced lines have been known for many years. They were described by F. E. Baxandall as satellite lines. The most reasonable interpretation is that they come from two streams of gas—one receding, as we observe from a point just preceding principal mid-eclipse, and the other rapidly approaching, as we observe from a point immediately after principal mid-eclipse. Before mid-eclipse the F-type star obscures the approaching stream. After mid-eclipse the F-type star obscures the receding stream. The two streams must therefore be located somewhere in the space between the two components.

It is significant that the spectral characteristics of these two sets of satellite lines are quite different from one another. The approaching stream is strong. Its lines are prominent and they correspond to a very high stage of ionization and excitation. They give the impression of a stream of gas of high temperature coming from a hot source. The receding stream which is observed just before mid-eclipse is characterized by much lower ionization and lower excitation, and gives the impression of coming from a source which is relatively cool. This is entirely reasonable. The approaching stream, seen after mid-eclipse, starts from the B9 star and moves in the general direction of the F-type star. The stream before mid-eclipse moves from the cooler F-type star in the direction of the hotter B9 star.

G. P. Kuiper has analyzed the mechanical behavior of matter in contact binaries, that is, in binaries which have a common gaseous envelope. He has shown that equilibrium conditions can be maintained only if within this envelope currents are set up such that a stream of gas leaves the more massive component and travels in the direction of the less massive component flowing along its following hemisphere. It can be shown from Kuiper's analysis that, if the velocities in the stream are large enough, this stream may leave the system altogether and form a ring of spiraling matter which gradually disappears into interstellar space. But it is likely that a part of the stream will not leave the system completely. Those atoms whose velocities are not very great will pass around the F-type star and return along its advancing hemisphere to the source of their origin in the B9 star.

Figure 31 shows the results of Kuiper's analysis. There is a marked similarity between his picture and that in Figure 30 constructed on the basis of the observations. The latter does not distinguish between a single ring and a succession of spiraling arms. We do not have observational evidence of those branches of the spiral which are very far removed from the source of origin.

The system of Beta Lyrae contains two stars, one of which is of type B9 and contributes 96 per cent of the light. This star is a giant, but its luminosity is lower than that of Beta Orionis. The other star is too faint to impress its light upon the spectrum, but its temperature must correspond to that of a star of class **F**. The semi-major axis of the B9 star is about twice that of the F star, and the separation between their centers is about twice the semi-major axis of the B9 star.

A stream of gas passes from the B9 star in the direction of the F star, along that side of the latter which becomes exposed to sight after central eclipse. These gases stream out with a velocity of about 300 km/sec, but the dispersion is large, and hence the violet satellite lines are broad. The gas is hot, coming from the surface of the B9 star, and the lines show some evidence of dilution, but less than the general ring-type structure throughout the rest of the cycle. The F-type component hides this stream from our view until phase 0.03 P, when quite suddenly it becomes visible in absorption. As the stream passes

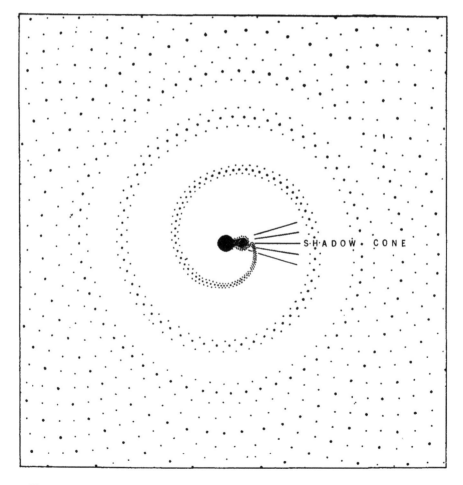

FIGURE 31.   G. P. Kuiper's theoretical representation of Beta Lyrae, in a frame rotating with the binary.

along the surface of the F-type star a part becomes deflected, through conservation of angular momentum, and forms a condensation at phase 0.1 P which has an average distance about twice the radius of the surface of the B9 star. This condensation continually receives matter from the outgoing stream of gas and dissipates it in the form of an expanding ring or spiral whose size increases, while its density decreases, as we pass from phase 0.1 P to phase 0.25 P and beyond.

But a part of the stream which gives rise to the violet satellite— perhaps that which consists of the slower atoms—flows around the F star and, after having been cooled along the surface of the latter while the exciting radiation of the B9 star is hidden by the bulk of the F component, returns to the B9 star as a stream having a velocity of about 200 km/sec. This stream gives rise to the red satellites. It becomes hidden from view at phase 0.02 P prior to central eclipse. That side of the B9 star which is subjected from above to the intense nebular radiation of hydrogen and helium from the outgoing stream and from the condensation at phase 0.1 P shows marked departures in the populations at those levels from which strong B9 absorption lines should normally originate. This latter point is of importance in interpreting the behavior of the lines of the B9 component. Those of its lines which are not complicated by emission coming from the nebulous mass about the system have a constant intensity throughout the 13-day cycle, but there are some strong lines of He I which undergo very marked changes. It is reasonable to suppose that these changes are somehow related to the presence of the emission lines, yet it is possible to show that these variations in absorption intensities cannot be explained simply as a result of mechanical overlapping of an emission feature, on the photographic plate, with an absorption feature. We must think of a physical cause, one that is due to the radiation of the nebula in precisely those frequencies which are excited in the reversing layer of the B9 star. It is only reasonable to assume that the populations in the various atomic states of the helium atom in the reversing layer of the B9 star will be influenced not only by the continuous radiation of the star but also by the discrete nebular radiation of the ring or spiral around it.

The emission lines come from the nebula, and much of their strengthening at primary and secondary eclipses is simply a result of the prolonged exposures which must be made at these times in order to obtain the spectrum of the system as a whole. We conclude from this that in all probability the intensity of the emission varies relatively little. At least it does not vary conspicuously during the two eclipses. It is difficult to make a detailed analysis of the radial velocities of these emission features. They are broad, and they are cut up by the various absorption lines of the B9 star or the B5 component and, at certain times, by the various satellites which appear in the spectrum. Nevertheless, as a first approximation we can say that the emission lines show a tendency to vary in radial velocity approximately in the manner that one would expect if the greatest concentration of nebular material is located at the condensation which we observe in the form of very strong absorption lines when the system is at phase 0.1 P.

The pinwheel of Beta Lyrae is about as large as the entire solar system. Since its distance is roughly 500 parsecs, even the most powerful telescope is unable to resolve its structure. The solar system lies almost exactly in the plane of the spiral, and it is because of this circumstance that we are able to explain the observations. If we were located at right angles to the plane of the spiral, Beta Lyrae would not vary in brightness and its spectrum would have normal absorption lines accompanied by a set of narrow emission lines, all undisplaced by velocity. Its spectrum would have been designated as of class Be, identifying it as one of a group of interesting but not very exciting stars. As a matter of fact, there are very few stars now known which resemble Beta Lyrae. One interesting example of this kind is RY Scuti, to which attention was first called by S. Gaposchkin. More recently the spectrum of this star has been investigated by D. M. Popper at the McDonald Observatory, and he has found that there is a substantial agreement between the results obtained in the case of Beta Lyrae and those obtained from the study of RY Scuti. Another example, HD 187399, has been investigated recently by P. W. Merrill.

The amount of matter which is lost each year in the expanding ring or spiral of Beta Lyrae is extremely minute, and the process can continue for hundreds of thousands or millions of years without appreciably depleting the mass of the components. Nevertheless it is probable that the masses of the two components will gradually approach equality. This is due to the gradual passage of nebulous matter from the B-type star to the F-type star and to the dissipation of some of this matter into interstellar space. If the two components should finally become equal in mass, the process discussed by Kuiper would cease operating and it is quite likely that this would happen in an interval of time which is short compared to the age of the stars. This may explain why there are so few stars now in existence which resemble Beta Lyrae. According to P. Guthnick the light curve of Beta Lyrae varies in an irregular manner, from season to season. He has attempted to explain these variations formally as changes in the parameter $k = R_1/R_2$ and has suggested that there is a transfer of mass between the components. The changes in the light curve are probably accompanied by irregular changes in the satellite lines and in the emission features.

However, it is not quite easy to understand why, among the hundreds of spectroscopic binaries, only three possess these peculiar features. It is true that Beta Lyrae is unusual in many other respects. According to Z. Kopal, its total mass is approximately 100 times that of the sun and thus it belongs to the most massive pairs on record. Yet there are several other systems which have comparable masses. For example, Plaskett's famous star, HD 47129, has a mass which is also of this order of magnitude. This star has a number of peculiar features, but they are not similar to those of Beta Lyrae. For example, Plaskett's star, though it has weak emission lines, fails to show the very peculiar and strong absorption features which are so characteristic of the ring spectrum of Beta Lyrae. Another peculiar characteristic of the system of Beta Lyrae is that the principal, brighter component, which is being eclipsed at primary minimum, is at the same time the larger and the heavier in the system. Ordinarily, those close binaries which consist of a hot star and a cool star have a small hot primary of large mass and a large cool secondary of small mass. This, for example, would be true of the Algol system.

Various workers have tried to explain the gradual change in the period of Beta Lyrae, which has a tendency to increase appreciably with time. Kopal and Kuiper have independently discussed this problem without having reached a definite conclusion, but in all probability the increase of period is somehow connected with the decrease in the mass of the system. Perhaps we can accept the observational fact that gas is streaming out of the system at all times as an indication that we have here an actual test of the evolutionary hypothesis which we shall advance. Beta Lyrae, according to our view, would be a very young system and one that has been formed only recently and is now in the process of separating to greater distances between the components and at the same time losing matter into interstellar space.

The next system which we shall consider is UX Monocerotis. The variation in the light of this star was first discovered in 1926 by Ida E. Woods at the Harvard Observatory. The period was determined soon afterwards by E. Hertzsprung as 5.9 days. The spectrum is A3 at maximum light. The photographic magnitude at maximum is 8.4; the primary range is 1.7 magnitudes and the secondary range is about 0.1 magnitude. The orbital elements of this system were first determined from the light-curve by Mrs. M. B. Shapley. According to her work, the total luminosity of the brighter component is about 0.8 of the whole luminosity of the system, while the secondary component accounts for only 0.2. The ratio of the surface brightnesses of the two stars is about 15, in keeping with the large difference between the two minima. The eclipse is almost central and the inclination is 90°. The radii of the two stars, expressed in terms of the radius of the relative orbit, are approximately 0.15 and 0.30. The average density of one star is only about 0.003 that of the sun, while that of the other component is about 0.09 that of the sun. The spectrum was first observed in detail by A. B. Wyse at the Lick Observatory. He described the brighter component as of type A5 and the fainter component, which he could observe only during the total eclipse, as G1 peculiar. The reason for this designation is that the Fe and H lines are weak, relative to the Ca I line 4227. Apparently Wyse observed no emission lines.

In December 1944, S. Gaposchkin of the Harvard Observatory undertook a spectrographic study of this system with the 82-inch telescope of the McDonald Observatory in Texas. The character of the spectrum is exceedingly peculiar. The hydrogen lines show emission components which vary in intensity with the phase. The absorption lines of hydrogen and Ca II undergo strange variations with phase and also from cycle to cycle. Gaposchkin collected a series of 32 spectrograms, which was not sufficient to disentangle all the remarkable changes which he had discovered. Hence, in 1947, a series of spectrograms consisting of 152 plates was obtained between January 22 and April 11.

Outside of eclipse, the spectrum shows complex absorption lines of hydrogen and ionized calcium whose intensities vary over a large range. There are also emission features of hydrogen, and probably of ionized calcium, of variable velocity and predominantly on the red side of the corresponding absorption lines. The spectrum shows a large number of faint lines of neutral and ionized iron, neutral and ionized calcium, ionized strontium, etc., all of which can occur in spectral classes A to G. These lines are often somewhat diffuse. Sometimes they are clearly double, especially near the elongations. One component is attributed to the G-type star, the other to the A-type star. During the total eclipse the lines of the G-type star are observed with great intensity. They show no ionized iron; presumably all ionized iron lines must be attributed to the A-type star alone. But even at principal minimum the lines of hydrogen and ionized calcium remain peculiar.

From the photometric data we can conclude that the eclipse is total because the light-curve shows a constant brightness at minimum. The spectra indicate that at principal minimum the G star is in front. The lines of the A star, especially those of ionized iron, are completely invisible during mid-eclipse. At the secondary eclipse the lines of ionized iron are strengthened; hence the A-type star is then in front. This strengthening of the lines of the A-type star is interesting, in view of the very small value of the depth of the secondary eclipse. According to Gaposchkin, in terms of light loss this secondary eclipse amounts to only 0.027—that is, only 3 per cent of the light of the system as a whole. It is difficult to understand how such a small change in the total light can produce so conspicuous a strengthening of the lines of ionized iron. It is probable that we are concerned here not only with the geometrical phenomenon of the secondary eclipse, but also with a real physical change in the intensities of the lines.

Considering all the relative intensities of the absorption lines in this system, we can conclude that, in the region of λ 4250, the luminosity of the G-type star is approximately equal to the luminosity of the A-type star. During the total eclipse the spectrum of the G-type star is observed without blending with the spectrum of the small A-type star. It is of great interest to observe the appearance of the lines of ionized calcium at this phase. They show a large negative velocity, of the order of −250 km/sec, which seems to persist throughout the time of totality, even though the amount of the displacement becomes smaller as the eclipse progresses. These absorption lines are broad and conspicuous, but their contours are more shallow than those of ordinary ionized calcium lines in single stars of spectral type G. On the red side of these broad absorption lines there are indications of faint and fairly broad emission lines. Apparently these emission features prevent us from seeing the normal undisplaced lines of ionized calcium in absorption coming from the reversing layer of the G-type star. Since the violet-displaced absorption lines that give a velocity of −250 km/sec are observed during the total eclipse when there are no features left belonging to the A-type star, we must conclude that the rapidly approaching ionized calcium atoms are located in front of the G-type star and stream towards the observer, along the line joining the two stars. During the total eclipse the motion of the G-type star is at right angles to the line of sight. Its normal lines would give a velocity equal to the velocity of the system as a whole.

The contours of these absorption lines present an interesting problem. They are approximately symmetrical, with an overall width of the order of about 10 Å. If the expansion were radial with respect to the center of the G-type star, without appreciable turbulence, we should observe an unsymmetrical absorption line in the case of a layer which is close to the G-type star's photosphere. Or, we should observe a broad emission line with a relatively narrow violet absorption border in the case of a very distant shell. The fact that we observe a symmetrical line, though a shallow one, suggests that the broadening is largely caused by turbulent motions. The shallow appearance of the line must be due to small optical depth of the expanding layer. It is tempting to explain this observation in terms of prominences coming from the G-type star. A field of prominences would be expected to show a large range of velocities. This would account not only for the large negative displacement but also for the great width of the absorption line. Moreover, the optical thickness of the layer might easily be quite small, so that the contour, though greatly broadened by the dispersion in the velocities of the prominences, would be shallow. The tendency of the emission lines to show relatively smaller velocities of expansion as the eclipse progresses might be taken to indicate a diminution in the tendency to produce eruptive prominences as we go away from the advancing hemisphere of the G-type star.

The expanding ionized calcium layer which is in front of the G-star during the principal eclipse does not give rise to hydrogen lines. We observe during totality only weak and narrow absorption lines of hydrogen on the violet borders of the emission lines. These absorption lines give velocities of the order of 50 km/sec of approach. There is no indication of any other absorption lines corresponding to a velocity of expansion of 250 km/sec. We must therefore conclude that the emitting gases which originate in the G-type star are poor in hydrogen but rich in ionized calcium and that, moreover, the hydrogen atoms are driven out with a much smaller velocity than are the atoms of ionized calcium. This does not imply that the abundance of hydrogen in the prominences of the G-type star is small. It probably means that the excitation of hydrogen atoms to the second energy-level is relatively small, and this is entirely consistent with what we know concerning the temperature of the G-type star. Incidentally, this phenomenon reminds one of the early theoretical work by E. A. Milne and others on the expulsion of calcium atoms from the solar reversing layer. The spectral types of the larger component of UX Monocerotis and of the sun are approximately the same.

But this is largely a phenomenon which we observe during the total eclipse and which does not tell us much about the advancing side of the G-type star. We observe the advancing side best at phase 0.75 P, that is, a quarter of a revolution before principal eclipse. At this time the G-type star is moving toward us, while the A-type star is moving away from us. At this stage we notice a very remarkable structure in the calcium lines. They show two components in absorption, one with a negative displacement of 300 or sometimes even 350 km/sec, and another corresponding to a relatively small velocity of recession. At first sight it is difficult to interpret these two components. The violet-displaced component, with a velocity of approach of 300 km/sec or so, is weak. The other component, displaced toward the red side of the spectrum, is strong.

The weak component, with its large velocity of approach at phase 0.75 P, is probably produced mostly in front of the G-type star, while the stronger component, with its relatively much smaller velocity of recession, is produced in front of the receding A-type star. It is not possible to be absolutely certain of this interpretation. What we observe in the spectrum are the two lines, one weak, the other strong. If we remember that the G-type star is red and that it diminishes rapidly in surface brightness as we go from the ordinary photographic region, at λ 4250, to the region of the Ca II lines at λ 3933, while the A-type star is blue and has a large amount of continuous radiation at λ 3933, then it is no longer unreasonable to suppose that what we observe is a large mass of calcium gas in front of the G-type star which approaches us rapidly, and a similar mass of calcium gas in front of the following hemisphere of the A-type star, which is receding with a much smaller velocity. If we accept this hypothesis it must be understood to represent mainly a statistical phenomenon. The prominences—or whatever we may call these streams of calcium atoms receding from the approaching side of the G-type star with a velocity which is many times greater than the velocity of the G-type star within its orbit—show an excess of velocities of approach over velocities of recession. Perhaps we must think of the prominences as being mostly able to escape from the G-type star with only a relatively small fraction falling back into the same hemisphere. Those prominences which leave the G-type star probably do not return to it, unless they do so after having completed an approximately circular path around the entire system. It is of interest to note that the Ca II lines observed in recession at phase 0.75 P have velocities which do not differ greatly from the velocity of the A-type star itself. One might think of them as simply streaming within the binary system, with velocities which are not very different from those of the A star, so that perhaps one might say that these atoms are not very greatly affected by the presence of this star.

When we observe the binary at phase 0.25 P, that is, a quarter of a revolution after principal minimum, the G-type star is going away from us. The A-type star is approaching. At this time we again observe weak Ca II lines which have a velocity of recession somewhat greater than the velocity of recession of the G-type star itself. These velocities, though exceeding the velocity of the G-type star, do not differ from it as much as do the velocities of approach which we observed at phase 0.75 P. Our interpretation is that we observe Ca II atoms after they have made a complete turn around the system as they fall back into the G-type star. At this same phase of 0.25 P we also observe strong absorption lines of Ca II with relatively small velocities of approach. These velocities are similar to the orbital velocity of the A-type star, and we can think of this stream as forming a continuation of the stream which produced velocities of recession at phase 0.75 P. Since the A-type star is bright at λ 3933, while the G-type star is faint, this would account for the difference in the intensities of the two components.

This picture is not as simple as it might at first sight appear. There are large differences in the intensities of the two components from one cycle to the next and sometimes even within the same cycle. There are also large differences in the radial velocities indicated by these double components of the calcium and hydrogen lines. These large and irregular changes in radial velocity first suggested that we were not dealing with any ordinary type of stellar motion in the binary system. In the absence of such irregularities it would have been natural to suppose that the components represent simply the motions of the centers of gravity of the two stars. The only way to account for the differences is to suppose that the prominence action is irregular in character and is undergoing rapid changes in time. These changes are so rapid that they can sometimes be observed in an interval of only six hours.

On this hypothesis, the observed component lines represent the integrated effect of absorption in a large mass of prominences which absorb mostly in Ca II and H. They are subject to macroscopic variations which manifest themselves as the integrated effect of a large number of microscopic variations. If this interpretation is correct we shall probably find that it is equally applicable to other cases of stream motion in close binaries and to the formation of expanding or stationary shells in rapidly rotating single stars and in other peculiar objects.

At the two elongations of UX Monocerotis, that is, at phases 0.25 P and 0.75 P, we also observe double lines of H. The component lines are relatively sharp; this is significant because it shows that the gas is not sufficiently dense to produce Stark broadening. We infer that when the two stars are observed side by side there are, somewhere in front of them, masses of prominences whose motions are predominantly those of approach and other masses whose motions are predominantly those of recession. In front of neither star are there any large masses whose motions are at right angles to the line of sight. If there were, such masses would produce strong undisplaced absorption lines. We do not observe such lines. There may, however, be a stream of gas that is not projected upon either star-disk and that would produce an undisplaced emission line. Such a line probably exists. The space between the two absorption components is definitely stronger than the continuous spectrum, and this emission, if that is what we see, presumably comes from those portions of the ring-like structure which are not projected upon the apparent disk of either component.

The proposed hypothesis, which is convincing in the case of the atoms of Ca II, also seems to give fairly satisfactory results in the case of the components of the hydrogen lines. The only difference which we must make is that the hydrogen atoms originate primarily in the advancing hemisphere of the A-type star and produce strong lines of approaching gases and weaker lines of receding gases. This is consistent with the observation made during the total eclipse, that hydrogen atoms are not excited to any great extent by the radiation of the G-type star.

The orbital motion of the star itself is inferred not from the components of the hydrogen and calcium lines but from the components of the faint metallic lines. These show a normal behavior and lead to a determination of the orbital elements of the system which is consistent with the photometric solution. The velocity of the system as a whole is 20 km/sec of approach. The orbital velocity of the G-type star is 60 km/sec, and the orbital velocity of the A-type star is 140 km/sec. The motions are approximately circular, and the masses of the two components are approximately four times that of the sun for the G-type star and one and one-half times that of the sun for the A-type star. The G-type star is the more massive, and the ratio of the masses is 2.3. The hydrogen lines and the lines of ionized calcium also have components which belong to the ordinary reversing layers of the two stellar bodies, but we cannot measure them because they are complicated by the sharp and strong absorption components produced in the streams.

The spectroscopic phenomena of UX Monocerotis, and especially the rapid variations in the intensities and velocities given by the shell lines of Ca II and H are unique among the eclipsing variables thus far observed. A photoelectric light curve obtained by Hiltner, Struve, and P. D. Jose in 1949–1950 shows that the star undergoes occasional outbursts in total light amounting to 0.1 or 0.2 mag which last a few hours. These are probably connected with the changes in spectrum.

Figure 32 represents a schematic picture of the system of UX Monocerotis. The existence of a tenuous ring of gas is common to Beta Lyrae and many other close binaries. The long arrows starting outward from the leading hemisphere of the G-type star represent the Ca II prominences which produce the greatly displaced absorption lines during primary eclipse and at phase 0.75 P. These arrows are much longer than the one representing the motion of the G-type star itself. The relative velocity of the outflowing prominences with respect to the surface of the G-type star is about 250 km/sec, but the corresponding lines are broad enough to allow a range of about 150 to 350 km/sec for the outflowing prominences. The mass of the G star is about $4\mathfrak{M}_\odot$ and its radius is about $7R_\odot$. Hence, the velocity of escape is

$$v_{\text{escape}} = v_{\text{escape}\ \odot} \sqrt{\mathfrak{M}/R} = 618\sqrt{\frac{\mathfrak{M}}{R}} = 470 \text{ km/sec.}$$

This is larger than the upper limit of 350 km/sec which we have observed, but the difference is not sufficient to prevent the prominences from curving around and returning by way of the path indicated in Figure 32. It should be remembered that the system of UX Monocerotis consists of a G-type giant with a less massive A-type companion, which is an unusual combination somewhat resembling the system of Beta Lyrae.

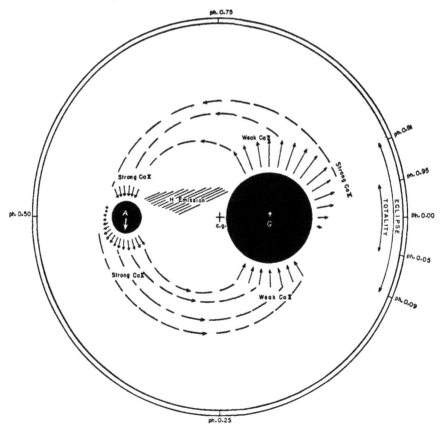

FIGURE 32. UX Monocerotis.

It is probably significant that in all of these cases we find evidence of a nebulous ring-like structure more or less filling the system of a close binary object. It is clear that some of the observations which were at first interpreted as rings around the hotter components could be interpreted equally well in terms of a structure somewhat like the one which we have pictured for UX Monocerotis. We are concerned with a phenomenon of great universality. Significant among the observational results are the distortions in the velocity-curves of such stars as U Cephei and SX Cassiopeiae. These two stars have light-curves which are entirely symmetrical. The secondary minima occur almost exactly half-way between the two principal minima. Hence, the quantity $e \cos \omega$ must be very near to zero. This does not necessarily mean that e is zero, but if we observe this sort of thing time after time in different stars we are justified in the conclusion that the great majority of eclipsing variables of very short period have almost circular orbits.

Yet the radial velocity-curves of both stars, U Cephei and SX Cassiopeiae, are very unsymmetrical. They have an asymmetry of the kind which would lead one to suspect that the eccentricity is of the order of 0.3 and that the longitude of the periastron ω is somewhere between 0° and 45°. It was believed at one time that perhaps the radial velocities of these stars were incorrect. The spectrum of U Cephei contains many broad and diffuse lines and is difficult to measure. Nevertheless, a study at the McDonald Observatory involving the measurements of more than 100 spectrograms leaves no doubt concerning the existence of this very remarkable asymmetry in the velocity-curve (Figure 19).

In 1930 E. F. Carpenter found $e = 0.47$ and $\omega = 25°$, so that

$$e \cos \omega = 0.43.$$

A new determination of the orbit in 1943 at the McDonald Observatory gave $e = 0.2$ and $\omega = 40°$, and

$$e \cos \omega = 0.15.$$

Although these values are different, they give a fairly good indication of the asymmetry of the velocity-curve. Adopting as the mean

$$e \cos \omega = 0.3,$$

we can show that this is not consistent with the fact that in the light-curve the shallow secondary minimum falls almost precisely half-way between two successive primary minima. The two mid-eclipses take place when $v + \omega = 90°$ or $270°$. Hence,

$$\gamma + Ke \cos \omega$$

is the radial velocity at these two moments. In the case of U Cephei, $\gamma = -5$ km/sec and $K_1 = 120$ km/sec, so that

$$v_{\text{eclipse}} = -5 + 36 = +31 \text{ km/sec.}$$

Referring to the velocity-curve in Figure 19, we find that the phases of the two minima should be:

> phase of primary minimum = 1.42 day
> phase of secondary minimum = 2.39 day.

The difference is 0.97 day, but the entire period is 2.49 days, and $1/2\,P = 1.25$ days. There is a difference of nearly 0.3 day, or 7 hours, between the spectrographic and the photometric intervals. Yet the photometric time of secondary minimum is certainly correct to within a small fraction of one hour. Unquestionably the evidence from the light-curve is correct, while that from the velocity-curve is incorrect, because the latter is distorted by absorption phenomena in a nebulous ring.

It is remarkable that a statistical study of the distribution of the longitudes of periastron in close binary systems observed with the stellar spectrograph has shown a strong tendency for ω to cluster around $0°$. This clustering is so pronounced that it has led to the

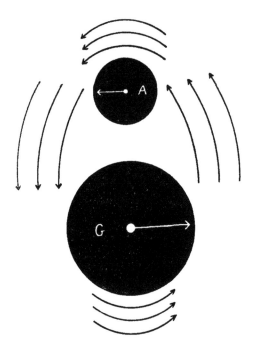

FIGURE 33. Gaseous streams in binary systems producing distortions of velocity-curves.

conviction that gaseous streams of the kind we have described in the system of UX Monocerotis are the rule rather than the exception.

Miss Elizabeth L. Scott has found that the clustering of ω near 0° is especially pronounced among binaries with periods from 2 to 5 days, and is not noticeable for periods shorter than 2 days. She noticed also that this phenomenon is present more strongly among binaries of early spectral class than among those of class later than about A5. The probability that the observed distribution of ω is due to chance is of the order of 0.0002 for the 87 systems whose periods lie between 2 and 5 days. This is a strong indication that distortions of velocity-curves are frequent among close spectroscopic binaries.

Figure 33 illustrates a system of streams which would account for the observed distortions of the velocity-curves. Outside total eclipse, we observe only the light of the A-type star. Just before the eclipse of this star by the G-type component begins, we observe a stream of gas projected upon the apparent disk of the A star. The radial velocity of this stream is one of recession, and it exceeds in amount the radial component of the orbital motion of the A star. Similarly, after the eclipse, an approaching stream of gas is observed in front of the A star. If the lines are blended, the result will be a distortion of the velocity-curve, causing a steep maximum just before the eclipse and an equally steep minimum immediately after the eclipse. But in reality the conditions of ionization of the two sides of the stream are not the same, and the blending may not produce equal distortions of the curve on both sides of the eclipse. It should be pointed out that there are many good examples of systems in which the spectroscopic eccentricities are real. To distinguish where an eccentricity of the orbit is real and where it is produced by intervening masses of gaseous streams is very difficult, but as a rule the Algol-type binaries which have small photometric ellipticities of the components are more likely to show the true values of e without serious interference from gaseous streams. Among the O-type binaries, Y Cygni is probably the best example of a system in which there is agreement between the photometric and spectroscopic eccentricities. In this system the ellipticity effect is small.

Another word of caution should be said with respect to the spectrographic observations. In any particular case it is difficult to distinguish between absorption lines produced in a stream and those produced in the reversing layer of a star. It is only in exceptional systems, such as UX Monocerotis, that we can immediately distinguish between the two. There are a number of systems of an intermediate character in which we observe with difficulty the existence of blended absorption lines. This, for example, is true in such a star as U Coronae Borealis. In still others, such as RZ Scuti, the velocity-curve itself presents a departure from the smooth run which it should have if it represented ordinary Keplerian motion. There is a conspicuous bulge in the velocity-curve just prior to the rotational disturbance which we have already described. This bulge occurs after the maximum of the velocity-curve has been reached at phase 0.75 P (or about 10 days), and just before the rotational disturbance becomes noticeable (Figure 34) at phase 14 days.

An excellent illustration of the effect of gaseous streams in binary systems is presented by SX Cassiopeiae. The variable light of this star was first discovered in 1907 by Mrs. L. Ceraski at Moscow. Numerous visual photometric observations were made by S. Enebo, M. L. Luizet, L. Pračka, and others. These were used by H. Shapley,

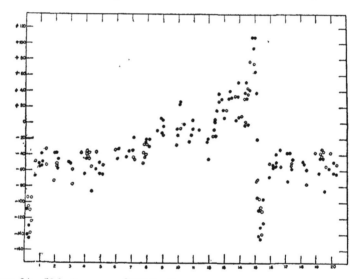

FIGURE 34.   Velocity-curve of RZ Scuti.

and later by Miss M. Fowler and B. W. Sitterly, for the determination of the photometric elements of the orbit of this eclipsing binary. In 1927, B. P. Gerasimovič published a light-curve which was based upon 905 estimates of the brightness, made on Harvard plates. The latest and most complete photometric study was made by R. S. Dugan in 1933. His paper contains a complete bibliography and a summary of previous photometric and spectrographic observations. The period, as determined by Dugan, is a little more than 36½ days. This is a rather long period for a star of the eclipsing type. The spectrum shows emission lines, and we have already seen that the presence of emission lines in binary systems is of considerable interest.

The visual magnitude at maximum is 9.0. The principal minimum has a depth of almost exactly 1 magnitude and the secondary minimum has a depth of about 1/3 magnitude. The eclipse is total, as is shown by the constant intensity of the light during the minimum (Figure 35), and the duration of the entire eclipse is 3.74 days. There is no noticeable departure in the position of the central point of secondary eclipse from the half-way mark between two successive primary minima. Hence it is concluded that the quantity $e \cos \omega = 0$. The light-curve has led to the derivation of the following photometric elements: the light of one star is 0.46 of the total, that of the other is 0.54. They are thus approximately of the same brightness. The ratio of the surface brightnesses however, is about 5, show-

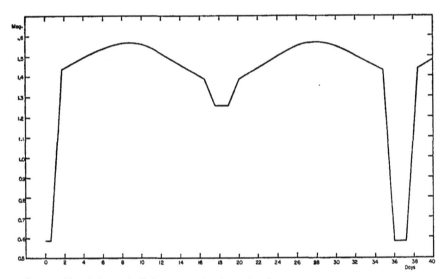

FIGURE 35.   Schematic light-curve of SX Cassiopeiae.

ing that the temperatures of the two components must be quite different. The radius of the brighter star, measured in terms of the distance between the two components, is 0.12 and the radius of the fainter component is 0.25. Thus we have here a typical Algol-type system in which the fainter component has the larger diameter and is a subgiant.

The spectrum of SX Cassiopeiae was observed by Adams and Joy at the Mount Wilson Observatory. The spectrum of only a single component was seen and was found to be A6s, where s designates that the absorption lines are sharp. During the eclipse, the spectrum is G6 which, of course, means that we observe only the light of the fainter component. According to the Mount Wilson observers, the A-type star resembles Alpha Cygni and is a supergiant. Dugan's values of the densities of the two stars suggest that they must be very luminous indeed. Joy states that at primary minimum, for about one day on each side of mid-eclipse, only the G6 spectrum is visible, without a trace of the A6 star. The A6 and G6 spectra are blended up to about two days before and after mid-eclipse. Beyond these limits Joy did not detect the spectrum of the late-type star. The light of the A star dominated. The H lines have bright fringes, and these emission features vary in intensity with the phase of the binary. The Mount Wilson observers state also that there have been "remarkable variations in the amplitude and eccentricity" for which no satisfactory explanation could be found.

Because of the presence of bright lines, this star was placed upon the observing program of the McDonald Observatory, and in the summer of 1943 forty-two spectrograms covering three consecutive

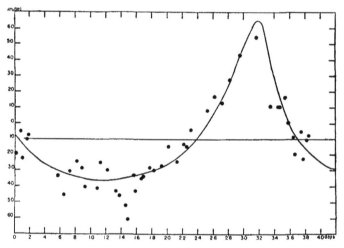

FIGURE 36.   Velocity-curve of SX Cassiopeiae.

cycles of this system were obtained. Figure 35 represents schematically the light-curve as published by Dugan. We see that this is a typical eclipsing light-curve with a considerable change in brightness outside eclipse, suggesting that the components are elliptical in shape. Figure 36 shows the observed radial velocities with an approximate curve drawn through them to illustrate the effect of the binary motion. It is at once evident that the two curves are not compatible with each other. The light-curve suggests that $e \cos \omega = 0$. The velocity-curve, on the other hand, suggests that the eccentricity is about 0.50 and the longitude of periastron $\omega = 10°$. Hence, the spectrographic elements give $e \cos \omega = 0.49$. This is such a large value that it could not possibly have escaped the attention of the photometric observers if it were real. It is, in fact, the quantity

$$e \cos \omega = \frac{\pi\left(t_2 - t_1 - \dfrac{P}{2}\right)}{P(1 + \cosec^2 i)},$$

where t_2 and t_1 are the times of the two eclipses, and i is the inclination. If we apply this formula, using the spectrographic determination $e \cos \omega = 0.49$, then we would find that the interval from principal minimum to secondary minimum should be $t_2 - t_1 = 29.7$ days, and the interval from secondary minimum to principal minimum should be $t_1 - t_2 = 6.9$ days. The theoretical formula for $t_2 - t_1$ is derived from the series expansion of $(v - M)$, the so-called equation of center. This derivation neglects powers of e higher than the first, and is only correct for small values of e. But we can derive $t_2 - t_1$ directly from the observed velocity-curve (as we have already done in the case of U Cephei) by remembering that

$$\gamma + Ke \cos \omega$$

is the radial velocity at the times of eclipse, when $v + \omega = 90°$ or $270°$. The result is not significantly different from the one given by the formula. Anything like the asymmetry of the two minima in the light-curve demanded by the spectrographic elements is out of the question. On the other hand, it is not possible to alter the spectrographic elements appreciably without violating the velocity-curve. For example, if we take $e = 0.37$, $\omega = 21°$, $K = 47$ km/sec and $\gamma = -10$ km/sec, we still find $e \cos \omega = +0.35$, and the discrepancy between the photometric and spectroscopic results remains conspicuous.

We must conclude that the measured radial velocities do not wholly represent the orbital motion of the brighter component. An appreciable portion of the measured velocities must come from some other type of motion, perhaps from a gaseous mass or ring which only in part belongs to the ordinary reversing layer of the A-type star. The dynamical properties of such a stream are not known. This hypothesis might provide an explanation of the peculiar variation in the eccentricity and amplitude of the velocity-curve, which had already been observed at the Mount Wilson Observatory by Adams and Joy. As a matter of fact, we now realize that most of these gaseous rings are unstable formations. They are not always present with equal strength. Sometimes they produce strong absorption lines, while at other times they produce weak absorption lines or even no absorption lines at all.

When this hypothesis was first proposed, it was distinctly revolutionary in character. If we cannot trust the velocity-curve of SX

Cassiopeiae and of U Cephei, for which the photometric observations provide the crucial test, how can we trust the velocity-curve of any other spectroscopic binary? And how much confidence can we place in the orbital elements derived from such velocity-curves? It is difficult to give a definitive answer to this question. The spectrum of SX Cassiopeiae is peculiar and, clearly, we should not trust a peculiar spectrum to give us the radial velocity of the center of mass of the star.

The emission lines show a most peculiar variation in intensity just before and immediately after the central eclipse. It turns out that in their normal condition these lines are present as two components. About one day before the beginning of the partial eclipse of the system—that is, about one day before the disk of the fainter G-type star begins to encroach upon the disk of the brighter A-type star—the emission lines begin to change in relative intensity. We notice a gradual strengthening of the red component at the expense of the violet component. This process continues for some time, until the violet component is about three or four times weaker than the red component. Thereafter, the two components become more nearly alike in intensity, and they reach equality at or very close to mid-eclipse. After mid-eclipse, the violet component becomes the stronger, and a short time afterwards it is five or six times stronger than the red component. This great intensity of the violet component continues until about one day after the end of the partial eclipse. The succession of the changes in the bright lines indicates an eclipse of the emitting gases, and is somewhat similar to that observed by Joy and others in RW Tauri (p. 196). We have, however, in SX Cassiopeiae no phase of total eclipse for the emission lines, but either a partial or an annular eclipse. And, of course, we cannot be certain that the emission lines originate in a rotating ring. The observations show only that about one day before the beginning of the photometric eclipse the G-type star begins to occult a mass of gas whose velocity is about -220 km/sec. This value is derived from the displacement of each of the emission components from the normal position of the line under consideration. At about the beginning of the total phase of the eclipse, or perhaps a little earlier, the violet component is the weakest, but it never completely disappears and is, therefore, never totally eclipsed. Similarly, at the end of totality the violet component is strongest, but the red component is still visible. Only a little later the red component is completely gone and is, therefore, at that time totally eclipsed by the G star. In the case of RW Tauri, Joy observed the disappearance of both components of the hydrogen emission lines. This indicated that the emitting ring of gas was smaller in diameter than the eclipsing star of that system. In the case of SX Cassiopeiae the emitting shell or ring is slightly larger than the

G-type star, and its diameter may be derived from the times when the eclipse of the bright lines begins and ends.

We must attribute the emission lines of SX Cassiopeiae to a stream of gas which flows towards us with a velocity of the order of 200 km/sec and is eclipsed before principal photometric eclipse. Since

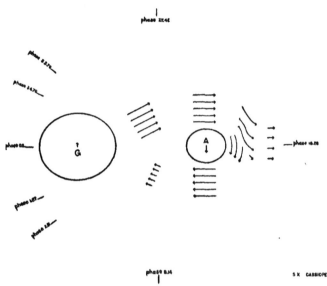

FIGURE 37. Schematic representation of the system of SX Cassiopeiae. The drawing is not quite to scale, the distance between the stars having been made slightly less than that given by R. S. Dugan. At phase 0.0 day we observe the center of the photometric eclipse. Beyond the G star there remain uneclipsed two streams of gas—one receding, the other approaching—which produce double emission lines of equal intensity. Immediately preceding partial eclipse, the receding stream is uneclipsed, while the approaching stream is covered by the G star. At the same time, a cool but rapid stream flowing from the G star to the A star is seen projected upon the disk of the latter, producing sharp absorption lines displaced toward the red. Immediately after the end of the photometric eclipse the approaching stream producing emission lines is uncovered and the receding stream is eclipsed. Upon the disk of the A star we now see projected a hot and slowly approaching stream which produces a shell spectrum with a small shift toward the violet. At phase 9.14 days the stars are at elongation. The shell-absorption lines show no relative shift, but the emission lines are produced from the stream at the far side of the A star, rushing away from the latter, and they fill in the central H absorption lines. At phase 18.28 days the emission lines are double, but the absorption lines are displaced toward the violet. We infer that the stream of gas rushing around the following side of the A star divides into two streams, as in Beta Lyrae. One part of the stream, presumably composed of the more distant or more rapid strata, expands outward, while the other part flows around the preceding side of the A star and ultimately returns to the G star. At phase 27.42 days conditions are similar to those observed at the first elongation: the velocity measured in the shell lines again equals the radial velocity of the A star, and the outrushing gases at secondary minimum cause undisplaced emission, which fills in the central hydrogen absorption lines. The arrows indicate the approximate directions of the motions of the stars and streams, but no attempt has been made to draw them to scale. Since the spectrum of the G star is observed only during totality and the slope of its velocity-curve during that interval is not sufficiently accurately determined, we have no knowledge about the relative masses.

the eclipse of the stream begins about one day earlier than the photometric eclipse, we attribute to the stream a height above the surface of the A-type star of the order of the latter's diameter. After photometric mid-eclipse we observe a stream of gas whose velocity is about 200 km/sec of recession. The entire picture is similar to that observed in Beta Lyrae, except that in SX Cassiopeiae the smaller and probably less massive star is the bright A-type component, while in Beta Lyrae the smaller and less massive star is the invisible F-type component. Figure 37 is a schematic representation of the system of SX

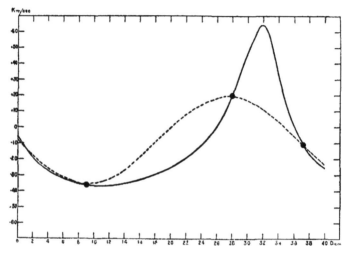

FIGURE 38.   Reconstruction of the velocity-curve of the A star in SX Cassiopeiae. The solid curve represents the observed radial velocities. In accordance with the model proposed in Figure 37, we assume that near mid-eclipse of the A star by the G star and at the two elongations (phases 9.1 and 27.4 days) the observed velocity is equal to that of the A star. This would indicate a symmetrical velocity-curve ($e = 0.0$), in harmony with the photometric results, which require $e \cos \omega = 0$. The broken curve has been computed under these assumptions, with $\gamma = -8$ km/sec and $K = 27$ km/sec. The large negative departures of the observed curve from the true curve are attributed to an expanding stream of gas from the A star, as seen between phases 10 and 27 days. The effect reaches maximum near secondary minimum. The large positive departures between 30 and 34 days are attributed to a cool receding stream of gas which flows from the G star toward the following side of the A star.

Cassiopeiae. It is more difficult to reconstruct the true velocity-curve from our observations. Of necessity we must make some assumptions as to whether we observe the real radial velocity of the A-type star at any stage within the cycle of the binary. Although there is a certain amount of arbitrariness in this interpretation, Figure 38 represents the best reconstruction of the velocity-curve that can be made at the present time. Three points in this curve have been chosen to represent the true velocity, and a sine curve has been drawn through these points. The solid curve represents the observed radial velocities.

If we now compare the dotted curve, which represents the true velocity variation of the A-type component, and the observed velocity-curve, which results from the measurements, we notice that between phase 10 days and phase 28 days the measured radial velocity is lower than the radial velocity of the center of gravity of the A-type star. On the other hand, between phase 28 days and phase 36 days the measured velocity is higher than the velocity of the center of gravity of the A-type star. This must mean that in the vicinity of secondary minimum we observe a mass of gas which is approaching us because it has a negative radial velocity. We must realize that what we actually observe is a blended line produced by the combination of a line in the reversing layer of the star with a line produced in the nebulous mass of gas. According to this picture, then, we should have an outflow of gas which approaches the observer when he sees the A-type star in front of the G-type star, and this will happen during the secondary minimum. This phenomenon bears a certain resemblance to the expansion of ionized calcium gas during the total eclipse in UX Monocerotis.

The large excess of the velocity-curve, as measured on our plates before primary minimum, compared with the normal sine curve which is consistent with the photometric orbit, has been found to constitute so common a feature of binaries of this type that we can regard it as a definitely established phenomenon. In some systems we observe a bulge in the velocity-curve for a short fraction of the period just prior to the beginning of the eclipse.

Adams and Joy pointed out long ago that outside eclipse the spectrum of SX Cassiopeiae is that of an A6 star resembling Alpha Cygni. The McDonald spectrograms confirm this description, but they add some interesting additional information. There can be no doubt that the spectrum is that of a very luminous A-type star in the later subdivisions of that class. But if the spectrum resembles Alpha Cygni, then where are the strong lines of Mg II 4481 and Si II 4128 and 4131? All three of these lines are present in SX Cassiopeiae, but they are surprisingly weak. Thus, the Mg II line is weaker than the Ti II line 4468 or the Ti II line 4501. In all normal supergiants of class A, and even in some F-type supergiants like Epsilon Aurigae, the Mg II line is stronger than these Ti II lines. This interesting weakness of the Mg II and Si II lines at once reminds us of the phenomenon of dilution which we have described in the case of Pleione. The conclusion is evident: we have before us at certain phases the spectrum of a shell whose dilution factor differs appreciably from 1. It is difficult to estimate the departure in the intensity of the Mg II line, but a factor of the order of 2 or 3 would probably be consistent with the observations. This, in turn, suggests that the shell or the ring is a formation which extends above the surface of the A-type star to a height of the order of one stellar radius. Although this conclusion is somewhat uncertain, it is in general agreement with the height which results from the eclipses of the emission components of the hydrogen lines.

The general character of the spectrum of SX Cassiopeiae shows a gradual change throughout the 36-day period. This is a somewhat unexpected result, because in ordinary spectroscopic binaries we do not often find such changes in the relative intensities of the absorption lines. In SX Cassiopeiae the intensities of the lines change quite gradually. First of all, there is a stage at which the lines which are commonly present and strong in expanded atmospheres or shells become very conspicuous, but there is also a tendency of the other lines, which presumably come from the reversing layer of the A-type star, to change appreciably with phase. For example, it is apparent that a few days before the beginning of the partial eclipse the lines of Fe II begin to fade, while the lines of Fe I increase in intensity. Thus the spectral type of the combined light could be said to become a little later even before the eclipse has started. This result is confusing because we expect that only during the eclipse the light of the G-type star would begin to predominate, producing a spectrum rich in lines of neutral Fe I, Ca I, etc., but deficient in ionized Fe II, etc. The fact that there is a change in spectral type prior to the beginning of the partial eclipse shows that we are not directly concerned with the light of the G-type star. This transition from a pure A-type shell spectrum to a later-type spectrum which finally, during the eclipse, becomes a pure spectrum of type G, suggests that the transformation is produced by the lines of the receding stream through which we observe the A-type star just before the eclipse begins. It is this stream which is responsible for the high peak in the velocity-curve. We must conclude that this receding stream has a later spectrum and presumably a lower temperature than the stream which we observe at secondary eclipse or prior to it. This state of affairs appears at first sight to be opposite to that observed in Beta Lyrae, where the stream starts at a higher temperature and where it flows out of the B8 star and returns at a cooler temperature, after it has flowed around the F star. But we must remember that in SX Cassiopeiae the roles of the hot and the cool star are reversed.

An unsolved puzzle is presented by the photographic light-curves of SX Cassiopeiae derived from Harvard plates by B. P. Gerasimovič and later by S. Gaposchkin. Both of these curves are entirely unlike Dugan's; they show no ellipticity and the A-type component is much smaller than the G-type component. Gaposchkin has attempted to explain this difference by assuming a semitransparent fringe which surrounds the A-type star and may even envelop the entire system.

But Russell has criticized this explanation (though not the observations upon which it is based), and for the present we may well accept his conclusion that this is an unsolved puzzle.

Only a few years ago we were confronted with a bewildering amount of observational information which did not make much sense and which could not be connected into a uniform theory. Fortunately the situation has changed completely during the past few months. We now possess a fairly satisfactory working hypothesis of the observed phenomena. The systems of close binaries are frequently enmeshed in nebulous matter which produces emission lines and absorption lines and which is responsible for some of the peculiar features observed in such systems as RW Tauri or SX Cassiopeiae. It now seems probable that the rings which we have described are intimately connected with the phenomenon of varying line-intensities in such systems as AO Cassiopeiae or Plaskett's massive star. The latter is known as HD 47129, and it was first investigated by J. S. Plaskett at the Victoria Observatory in 1922. On December 16, 1921, he found double lines in its spectrum, both components belonging to spectral type O. From the measurements of 30 spectrograms he found that the period is about 14.5 days, while the semi-amplitude of the velocity-curve of the primary is 206 km/sec and that of the secondary 247 km/sec. These large velocity ranges, combined with the relatively long period, result in very large masses for the two components. Plaskett obtained 76 times the mass of the sun and 63 times the mass of the sun, for the quantities $\mathfrak{M}_1 \sin^3 i$ and $\mathfrak{M}_2 \sin^3 i$, respectively. In the winter of 1947 to 1948 a new series of spectrograms of this remarkable system was obtained at the McDonald Observatory. The primary gave velocities in accordance with Plaskett's original orbit, but the secondary turned out to be entirely anomalous. When the secondary component is receding, its intensity is extremely faint; when it is approaching, its intensity is relatively strong. On some plates the secondary appears almost as strong as the primary. This is especially true in the case of helium. It is much less pronounced in the case of hydrogen. This phenomenon closely resembles in character that which we have already described (p. 182) for Alpha Virginis, Mu-one Scorpii, V Puppis, Sigma Aquilae, Beta Scorpii, and the W Ursae Majoris systems.

Even more interesting is a comparison of the results obtained in two different cycles of this binary. It turns out that both the intensities and the displacements of the secondary components are subject to large erratic variations at identical phases in two different cycles. The secondary can be 100 km/sec or so, different in radial velocity, and its intensity can also differ by a factor of two or three (Plate XVII). It is difficult to explain these peculiar variations. Clearly, they cannot be caused by the ordinary reversing layer of a normal star. We must be observing not a simple star which can be treated as a geometrical point, but a mass of gas which is only loosely connected with the stellar body. In the light of all the other observations which have accumulated, it is reasonable to suppose that the secondary, and perhaps also the primary, are subject to a large amount of prominence activity. These prominences presumably leave the advancing side of each component with velocities which are sufficient to throw them entirely out of the gravitational domain of each star. We do not know what the motions of these gaseous masses will be,

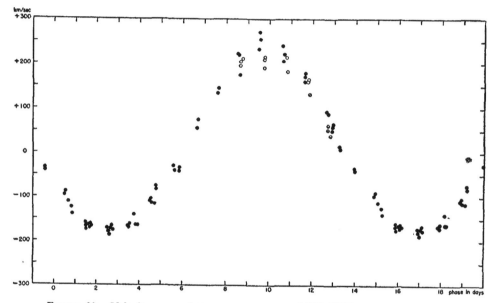

FIGURE 39. Velocity-curve of stronger component of HD 47129.

but it seems quite likely that the activity is not always the same. It varies in intensity, thus producing lines of different strength; and it varies in velocity, thus producing variable displacements of the spectral lines. It is somewhat difficult to understand why this prominence activity appears to be so much more concentrated in the secondary component than in the primary, but we have already seen that in all other systems with prominence activity we are primarily concerned with the spectrum of the secondary star. In the case of HD 47129 there are some indications that the radial velocities of the primary do not always exactly repeat themselves; we conclude that the prominences probably influence its lines also.

The radial velocities of the two components of HD 47129 are shown in Figures 39 and 40. The fainter component shows a striking difference of its average velocity from that of the stronger component. This suggests that the gases surrounding this component expand with an average velocity of the order of 100 km/sec. A somewhat similar system is AO Cassiopeiae. It is also known as Boss 46, and its orbit was determined at the Mount Wilson Observatory by Adams and Stromberg. The period is almost exactly three and one-half days and the spectral type is O. In 1926 the orbit was determined again by J. A. Pearce at Victoria. He discussed his spectrographic results in conjunction with the light-curve of this system derived with a photoelectric photometer by P. Guthnick at Berlin-Babelsberg and

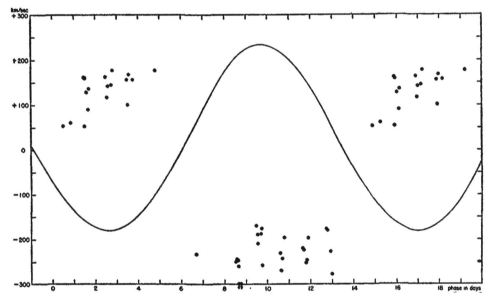

FIGURE 40.   Velocities of fainter component of HD 47129 showing a general tendency to give large negative values.

found that the brighter component is 36 times more massive than the sun, while the fainter component is 34 times more massive than the sun. The spectrum resembles very closely that of Plaskett's star, HD 47129. New McDonald observations by Struve and H. Horak show that the secondary component is also subject to variations in intensity. It is weak when the secondary is receding and it is strong when the secondary is approaching. There may also be changes from cycle to cycle.

The velocity range of the fainter component is smaller than that of the stronger component, which, if interpreted in the usual manner, would mean that $\mathfrak{M}_1/\mathfrak{M}_2 > 1$, and that the individual masses are smaller than those derived by Pearce. A photoelectric light curve by

Hiltner shows a much sharper minimum when the secondary star is eclipsed than when the primary star is eclipsed, and there is a general similarity to the light curve of the Wolf-Rayet binary CQ Cephei which suggests that electron scattering may be important in both objects.

10. Summary of Observational Results on Spectroscopic Binaries

I. Variations of line-intensities seem to be a fairly common phenomenon in close binaries. They have been observed in all W Ursae Majoris systems in which the observations are sufficiently precise. They have also been observed in the early-type systems V Puppis, Mu-one Scorpii, Alpha Virginis, Sigma Aquilae, Beta Scorpii, HD 47129 and AO Cassiopeiae. They are probably variable in amount, being very conspicuous at certain times and less conspicuous at others. This accounts for the failure of some observers to detect them in systems in which they were certainly present at an earlier period.

II. The second set of data refer to observations of general expansion as measured by the lines of the secondary components in such stars as HD 47129 and AO Cassiopeiae. There has been some evidence of unequal gamma velocities as determined for the two components of some spectroscopic binaries. In the case of HD 47129 the secondary shows a radial velocity which is about 100 km/sec smaller than the velocity of the system as measured by the brighter component. This suggests prominence activity on a very large scale, principally from the advancing hemisphere of the fainter component. We have no observational data to show whether there is a small amount of expansion from the brighter component, but perhaps it is plausible to think of an unsymmetrical envelope as being produced by prominences which are ejected from the advancing sides of both components with velocities that exceed the orbital velocity and that may occasionally exceed the velocity of escape. The prominence interpretation is rendered probable by the erratic variations in the intensities and displacements of the lines of the secondary components in HD 47129 and AO Cassiopeiae.

III. The most convincing example of prominence activity primarily associated with the advancing hemisphere of each star was observed in UX Monocerotis, where the velocity of the Ca II atoms ejected by the G-type star is, on the average, 250 km/sec with respect to the surface of the star. We do not know accurately the velocity of escape for this object, but we can estimate that because of its large size and its relatively small mass the velocity of escape will be considerably smaller than the value of 618 km/sec which applies to the sun.

But this quantity is probably not particularly important, because observations of expanding atmospheres strongly suggest that the pressure of radiation greatly exceeds the gravitational attraction for atoms which have been raised high above the normal reversing layer and are capable of absorbing spectral lines in those parts of the star's continuous spectrum in which the radiation is strong. We have never observed expansion in W Ursae Majoris systems, but it is not unreasonable to suppose that their unsymmetrical envelopes are unstable and disintegrate into space. The radial motions involved are probably too small to be measured with the spectrographs of small dispersion which we are compelled to employ in the study of these objects.

IV. The most complete information involving streams of gas passing from the more massive to the less massive star and back again, together with a general nebulosity around the whole system which expands in all directions, was observed in the case of Beta Lyrae and, more recently, RY Scuti. The latter star has been only superficially investigated at the McDonald Observatory, but it appears probable that it has a shell in which strong absorption lines are produced, and that the observed motions represent the blending of lines in the shell and in the stellar part of the system. In Beta Lyrae, as in all the other peculiar objects, the streams of gas move in the same direction as do the two stars.

V. In a group of eclipsing variables in which one component is a star of type A or B, variable emission lines have been observed which indicate the presence of a stream of gas surrounding the hot component of the system and revolving in the same direction in which the system is rotating. The first star to show this effect was RW Tauri. It has also been found in numerous other variables of the eclipsing type, including SX Cassiopeiae, RX Cassiopeiae, RW Persei, and a number of others. Apparently these streams are a common occurrence in spectroscopic binaries, but they are best seen during the partial phases of the eclipse, when the total light of the system is diminished and one side of the revolving ring is eclipsed. A. H. Joy, who discovered this phenomenon in RW Tauri, explained it in terms of a gaseous ring around the A-type component, but the observations do not really show whether we are concerned with a ring around the A star or with a stream of gas passing from one component to the other, and back again along the other side. The important thing is that the stream is revolving rapidly in the direction of the orbital motion.

VI. In several spectroscopic binaries there are definite indications that their velocity-curves are distorted by absorbing gases produced in rotating streams. For example, in U Cephei and in SX Cassiopeiae the photometric observations lead us to the conclusion that the eccentricity is negligible; yet the velocity-curve is unsymmetrical, resembling one whose eccentricity is about 0.3 and whose longitude of the periastron is between 0° and 45°. A closer study of the spectra of these stars has shown that the lines are not simple. Near the maximum of the velocity-curve they look as though they were composed of two components. Apparently, what we measure is the blend of one line which is produced in the normal reversing layer of the star, and another which is produced in a rapid stream of gas which recedes at the time of observation, and which precedes the time of mid-eclipse by about 0.1 P or 0.2 P. Such a stream would be similar to those whose existence we have observed in Beta Lyrae, but in this case it would be a stream of gas starting from the advancing hemisphere of the fainter component and moving toward the receding hemisphere of the stronger component. There is a similar stream starting from the advancing hemisphere of the brighter component and ending at the receding hemisphere of the fainter component. The absorbing properties of this approaching stream, which we might expect to observe immediately after mid-eclipse, are less pronounced than those of the receding stream. This difference may be due to ionization. In the star RZ Scuti the velocity-curve is not smooth and we can actually see two distinct maxima: one in the normal position, at phase 0.75 P, and the other in the displaced position, at phase 0.85 P.

VII. Our next group of observations refers to the velocity-curves of spectroscopic binaries. Statistically, there is a predominance, among systems having periods of the order of 2 to 5 days and spectra of types B and A, to give velocity-curves which are unsymmetrical in such a way that if they are explained in terms of orbital motion the eccentricity is large and the longitude of the periastron falls in the vicinity of 0°. A statistical study of the entire material of more than 500 velocity-curves shows that this unsymmetrical distribution of the longitudes of periastron cannot be the result of chance. It is probable that we have here a distortion of the velocity-curves, due to the blending of the normal stellar absorption lines with lines produced in a rapidly rotating stream of gas whose direction is that of the orbital motion but whose velocity is much greater. The approaching streams must be weaker than the receding streams, either because the density is greater or because the ionization and the excitation are less favorable. This produces a false maximum of the velocity-curve, shifted slightly towards later times, so that instead of observing it, in the case of a circular orbit, at phase 0.75 P, we observe it at 0.85 P, or even 0.90 P. As a rule, the minimum of the velocity-curve is much less affected.

Conclusions

The observational evidence has been accumulated over a period of many years, and there have been gradual changes in our interpretation of the results. There are, however, certain conclusions which can be drawn from the entire volume of results. Close binary systems are usually embedded in rapidly rotating gaseous envelopes which move in the same direction as the stellar components but, on the average, have larger velocities. The motions of this envelope are not circular, but they follow a somewhat complicated path, illustrated in Figure 28. The envelope is not symmetrical with respect to the stars. It looks as though it were shifted forward in phase along the direction of the orbital motion, so that there is more gas in front of the advancing side of each star. This envelope is usually not stable, but it disintegrates into space with velocities of expansion which are sometimes of the order of 100 km/sec, but are in most cases much smaller. The ionization and the excitation of the gases in these envelopes differ from star to star and also from one side of the stream to the other. The radiation of the stellar components probably accounts for the ionization of the gases in the envelope. It is not now possible to say how common this phenomenon may be. It is probably always present in systems of small separation between the components. It is not only present in systems whose components are main-sequence stars, but also in some in which one or both components are giants. The envelopes produce marked effects of absorption, and sometimes of emission, but so far as we can determine now they do not appreciably influence the continuous radiation of the whole system, except in the W Ursae Majoris stars and in a few peculiar early-type binaries such as Beta Lyrae.

The density of the envelopes is small, certainly much smaller than that of a normal stellar reversing layer. In the W Ursae Majoris systems we have attributed to these envelopes much of the radiation of their fainter stellar components. This would presuppose a considerable thickness and density, at least in the deeper layers of the shell. Thus, the total mass of the envelope may also be larger than that estimated from spectroscopic data in early-type binaries, namely $10^{-8}\mathfrak{M}_{\odot}$. But the extent of the envelope is much smaller than that envisaged by von Weizsaecker. There remains unanswered a rather curious question: Why are there several early-type systems in which the luminosities of the fainter components depart from the mass-luminosity relation in much the same way as do the fainter components of the W Ursae Majoris systems (p. 26)? Is this an entirely different phenomenon, or does it also mean that a common envelope renders the fainter components more luminous than is consistent with their small masses? In these systems the spectral types of the fainter components are, however, much later than those of the primary components. Hence it is probable that the departures from the mass-luminosity curve are not caused by common envelopes, but by some other process, such as the one described on page 248.

11. The Evolution of Close Double Stars

It is interesting to speculate on the possible course of evolution of a close binary system. The problem is very involved, and the theory of rotating single stars is in its infancy, not to speak of close binaries surrounded by common envelopes. At this stage we cannot formulate a definitive hypothesis, but we can discuss the existing data of observation from the point of view of stellar evolution. Even if the picture thus obtained may not suffice to reconcile different lines of thought, it will serve as a useful tool for research and stimulate further investigations.

Let us consider the W Ursae Majoris systems. We have already seen that they revolve very rapidly around one another. For example, in the case of AH Virginis the orbital velocity of the more massive component is 105 km/sec, while the orbital velocity of the fainter component is 250 km/sec. In each case the velocity is measured with respect to the center of gravity of the system. Because of the broadening of the spectral lines, we believe that the periods of rotation are equal to the period of orbital revolution. What, then, will be the effect of the common envelope created, as we suppose, by prominences which are formed predominantly on the advancing side of each star?

Undoubtedly the tendency will be a gradual loss of orbital momentum by the two stellar components. This orbital momentum will first be transferred to the prominences which escape from the stars and form a ring around them. The rings are unstable formations. As in the case of Beta Lyrae, they must dissipate into space and, in doing so, they must carry with them the orbital momentum which they have taken away from the stars.

Loss of orbital momentum may produce a spiraling inward of the two components, with a consequent decrease in the distance between them. But we must distinguish here two different effects. The prominences which erupt from the advancing side of each star reduce the masses. If there were only a reduction in mass without a consequent change in the orbital momentum, the result would be an increase in the distance between the two stars, in accordance with the computations by J. H. Jeans, W. D. MacMillan, E. W. Brown, and others. But in this case we are concerned also with a change in the angular momentum. It is as though we were shooting off projectiles from the advancing side of a large mass. The projectile carries with it a certain amount of the momentum which was originally contained in the combined mass, and this results in a diminution of the momentum of the remaining stellar object. But our data are not sufficient to predict, from theoretical considerations alone, whether the result will be a spiraling inward or an increase in the separation. All we are certain of is that such a binary will change in character; it will evolve.

Let us consider, for definiteness, a W Ursae Majoris system of spectral type K0 or late G and consisting of two components with masses of the order of $1\mathfrak{M}_\odot$ and $\frac{1}{2}\mathfrak{M}_\odot$, respectively. These two components revolve around one another almost in contact, inside a common envelope which is unsymmetrical, being thicker in the direction of the motion of each star. The erratic variations in the period, the light-curves and the velocity-curves of these systems, as well as the dynamical considerations we have just mentioned, lead us to conclude that this system is not permanently stable, but must change as the result of inner forces and without resorting to perturbations from other stars.

We do not know whether the components will tend to coalesce and form a single system or whether they are separating, having in the past been a single star. Figure 41 illustrates the possible evolutionary paths which a binary system of the W Ursae Majoris type may follow in the future and which it may have followed in the past. Let us suppose that the double star tends to draw together and that it will ultimately form a single object. If this process should occur without loss in mass, we should expect to find a descendant star of 1.5 solar masses and of angular momentum equal to the total angular momentum of the parent binary. This would result in such a large rotational velocity for the descendant star that we would easily observe it. The spectral type of the descendant body would be about G0, or perhaps late F, provided it remained on the main sequence. If it became disturbed in the process of coalescence to such an extent as to move away from the main sequence, it would be located in a part of the H-R diagram in which there are very few stars; it could not remain there very long and would presumably soon return to the proper place in the main sequence, which corresponds to 1.5 solar masses.

But we have never observed single stars of the appropriate spectral type and mass with rotational velocities sufficient to account for the angular momentum of the original binary. Hence, we conclude that the path designated in the diagram as A is not possible.

We explore next the possibility that the process of coalescence is catastrophic in character and involves the creation of a single star of

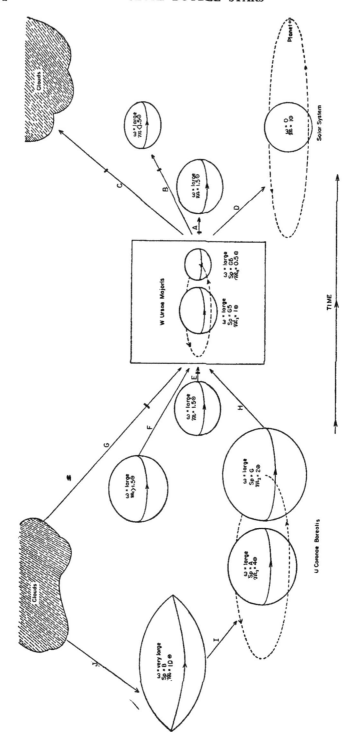

FIGURE 41.   Evolutionary processes in close double stars.   Improbable processes have been indicated by a pair of parallel lines.   The most probable course begins with a cloud of interstellar material, followed by the processes J, I, H, D.

less than 1.5 solar masses, because the remainder would be carried off into space. Unless the process of dissipation of the excess mass is quite peculiar, the resulting single star would have a tremendous velocity of rotation, and we have already seen that there are not now in existence in our galaxy single stars that could be regarded as the descendants of the W Ursae Majoris binaries. This eliminates path B. It is conceivable that the catastrophe is so tremendous that the entire mass is converted into a cloud; the star disappears from the H-R diagram and the cloud becomes endowed with the necessary angular momentum. This process is improbable, because from all appearances the W Ursae Majoris stars are even now in the process of change between a single body and a binary. There is no reason why the process should become explosive in character, nor have we ever observed among the W Ursae Majoris stars any indications of explosions such as we observe in certain O-type binary systems or in peculiar systems of the kind of Boss 1985 and others. We cannot definitely exclude the evolutionary path C, but we consider it highly improbable.

Before we discuss path D, let us turn to the opposite side of the diagram. If the binary resulted from the fission of a single star having the combined mass of the W Ursae Majoris binary, this parent body would have a spectral type of approximately G0, or late F, and since there are not now in existence single stars possessing the required spectral type and mass, together with the required angular momentum, we exclude the evolutionary process designated as E.

Path F cannot be easily discarded, because in this case we suppose that the parent star had a mass larger than 1.5 times that of the sun and that, in the process of fission, part of the mass was lost and was permitted to escape into space. There are plenty of single stars of masses larger than about ten times that of the sun, which possess the required amount of angular momentum. If, in the process of fission, about 90 per cent of the mass was evaporated into space and only 10 per cent was retained to form the binary, then path F would appear to be possible. But it is not a probable path, because we do not expect that in the course of an explosion in which most of the mass of the star is blown off into space, a binary can be formed which possesses such a remarkable similarity, in every case, to the typical W Ursae Majoris system. For example, the spectral types of the components, the ranges of the light-curves, the periods, and the character of the velocity-curves are remarkably similar in all the systems of this type. It does not seem reasonable to suppose that the explosion would always divide the remaining 10 per cent of the mass in precisely the same way. Hence we consider path F as improbable.

Path G, which is the reverse of C, is also improbable, because we should not expect that through the condensation of separate nuclei in a cloud of interstellar dust a binary will be formed that has always approximately the same physical and dynamical properties.

We are apparently left with only path H to represent the possible past history of our W Ursae Majoris binary, and path D to represent its future course of development.

In process H, most of the mass of the early-type binary is lost, but this loss can be explained in terms of the unstable common envelope which is so characteristic a feature not only of the W Ursae Majoris binaries but also of many binaries of early spectral type. In the process of dissipation of such a nebulous envelope, forces are at work which have precisely the same significance as those suggested by von Weizsaecker and by Ter Haar in their theories of evolution of single stars by rotational disintegration.

If we adopt path H, we next inquire whether there is any difficulty in explaining the origin of an early-type binary of large mass, such as U Coronae Borealis. It is evident that the difficulty of angular momentum does not exist. There are plenty of single stars having masses equal to the sum of the two components of U Coronae Borealis, and having at the same time a rotational velocity close to the value of rotational instability. Hence, we can easily find in the galaxy any number of objects which could be regarded as the parent stars or, if we wish, the descendant stars of our early-type binaries. The only reason for choosing path I is that we have already expressed a preference for path H as representing the future development of the early-type binary and, at the same time, the past history of the W Ursae Majoris binary.

Adopting this scheme for the past history of W Ursae Majoris, we are left with only one plausible course for its future development. This is indicated by path D, illustrating the formation of a planetary system with a single star having approximately the mass of the sun and devoid of appreciable angular momentum, surrounded by one or more planets at considerable distances, which contain a considerable fraction of the original momentum. The remainder of the original mass is supposed to have escaped to infinity, carrying off at the same time a large fraction of the angular momentum.

This is in effect the von Weizsaecker process of cosmogony which he first applied to the problem of the origin of the solar system and which he later expanded to include the development of single stars. The nebulosity in which turbulent motions are set up would be the common envelope surrounding the two components.

The outline of evolution which we have suggested is the following: Through condensation of an interstellar cloud, a young single star of early spectral type is formed (path J). As a rule, this star possesses a large amount of angular momentum. As it condenses, a stage of rotational instability will set in. It will shed matter at the equator and will thus lose some mass and orbital momentum. By a process of explosion it may divide into two components, forming a binary of early spectral type, like U Coronae Borealis (path I). This binary forms a common envelope and loses mass through its disintegration. A rough estimate for the mass of the envelope is $10^{-8}\mathfrak{M}_\odot$. Hence, if this dissipation requires the renewal of the gases in the envelope once every year, the process H would require 10^8 years. This would correspond to an average velocity of expansion of only 0.02 km/sec, which cannot be measured by the Doppler effect. In reality, we have observed expansions of the order of 100 km/sec in Plaskett's star, and of 75 km/sec in Beta Lyrae. These velocities are extreme values, and may not last throughout the entire cycle of evolution. Perhaps an expansion of 2 km/sec would be a reasonable guess. This would reduce the time-interval required for process H to 10^6 years, a value which agrees exactly with von Weizsaecker's estimate required for turbulence to produce a similar change in his hypothetical single star surrounded by a nebula (p. 146).

But a longer time scale, of the order of 10^8 or 10^9 years, is preferable on other grounds. The average motions and the distribution of the W Ursae Majoris binaries resemble those of single stars of the corresponding total mass. Hence, we must allow enough time for a redistribution of these quantities to be accomplished through the action of encounters and close passages, and this cannot be done even in 10^9 years unless the density of the system were much greater than it is at the present time. Moreover the average period of a system like U Coronae Borealis is several days, while that of a W Ursae Majoris system is 0.3 day. Hence, if T is the time required for process H to operate, we must have an average change

$$\Delta P \approx \frac{1}{T} \text{ day/year.}$$

The observations of hundreds of early-type eclipsing variables show fluctuations in P—some increasing, others decreasing—but there is as yet no conclusive evidence that there is a gradual diminution. If we can determine the epoch of eclipse in a binary whose $P = 1$ day, within 0.01 day, then in about 30 years the uncertainty in P would be about 10^{-6} day, so that we should just be able to observe the change if $T = 10^6$ years. In a number of binaries we can determine the time of mid-eclipse considerably more accurately, and for these stars the argument becomes even more critical.

As G. Kron has shown, there is reason to believe that in some systems small irregular changes in the distribution of light on the surfaces of the two stars may account for the erratic observed changes in P. But F. B. Wood has found that nearly all close binaries with well-established irregular changes in period are at the same time close to instability in the sense that their less massive components almost completely fill the volume of space available to them in the limiting Jacobian equipotential surface which has a common double point between the two components. If $\alpha = \mathfrak{M}_1/\mathfrak{M}_2$ is known, it is possible to compute the maximum possible value of b_f in units of the distance between the centers. For example, the light curve of TW Draconis shows a large ellipticity effect between the eclipses, and the shortest equatorial radius of the fainter star $b_f = 0.289$. The equation of the Jacobian surfaces shows that if $\mathfrak{M}_1/\mathfrak{M}_2 = 2.5$ the volume of the smaller loop is completely filled, and if $\mathfrak{M}_1/\mathfrak{M}_2 > 2.5$ a portion of the mass falls outside the limiting surface, and must escape. The spectrographic observations by Pearce give $\mathfrak{M}_1/\mathfrak{M}_2 = 3.6$. Wood finds that small disturbances, in the form of prominences, from the leading or following hemispheres of each star can account for the observed increases or decreases in period. For a hypothetical system having $\mathfrak{M}_1 + \mathfrak{M}_2 = 2\mathfrak{M}_\odot$ and $P = 2$ days the loss of mass needed to produce a change in P of one second is $10^{-6}\mathfrak{M}_\odot$. A change in mass of $10^{-8}\mathfrak{M}_\odot$ would be consistent with some of the observations.

The next process is D, which, in effect, represents von Weizsaecker's cosmogony. Only, here we think of the common envelope of a W Ursae Majoris system as the medium in which turbulent cells are set up which carry off mass and angular momentum. In doing so they may, again in accordance with von Weizsaecker, produce a system of planets. But the formation of planets is not an essential feature of our hypothesis. We have introduced it only because the common envelope of a close double star represents the sort of medium required for their origin, and because the sun is a typical, cool, slowly rotating dwarf of the kind we have described as the descendants of the W Ursae Majoris binaries.

We have already discussed process J, but we have not yet given an explanation of process I, the division of a single rapidly rotating star into a close binary. The classical fission theory of G. Darwin and of J. H. Jeans has for many years dominated the field. We have seen that, according to the latter, a rapidly rotating star may get rid of its excess momentum through the creation of a double star, as well as through the shedding of matter at the equator. We have shown elsewhere (p. 231) that there is observational evidence of the gradual shedding of gaseous matter by O-type and B-type stars in the form of ring-like structures which give rise to the phenomenon of emission lines and which sometimes create peculiar absorption spectra when we observe the continuous light of the star through the gases of the ring. We are much less certain just what takes place when a star divides and forms a binary. We do not observe such a process at work. We only observe the ultimate product after the binary has been formed. It is true that some binaries, like Beta Lyrae, strongly suggest that they have been formed quite recently. Yet we do not have any record of the transition of such an object from the single-star state to the binary star.

The most serious objection which has been raised against the classical fission theory rests upon the dynamical arguments of F. R. Moulton, W. D. MacMillan, and others, who have shown that a binary system cannot be produced by the splitting of a single star unless the original density was much smaller than that usually found in stars. Another serious argument against the fission theory is the recognition that ordinary processes of stellar encounters and, even more so, internal processes produced by tides and similar mechanisms, are not sufficient to produce the wide pairs which we observe visually. Several workers have attempted to evade this difficulty by assuming that the wider visual pairs have originated in some other manner than have the closer spectroscopic and eclipsing pairs. However, there are strong arguments against such an artificial division of the binary systems into two groups. In a series of papers in the Publications of the Astronomical Society of the Pacific in 1935, G. P. Kuiper has presented several arguments supporting the idea that all binary systems are essentially of one kind. For example, he found that the frequency curves of Δm, which represent the differences in the magnitudes of the two components of a double star, are nearly the same for the visual and spectroscopic pairs of the same absolute magnitude and spectral class. There is a very close relation between Δm and the ratio $\mathfrak{M}_2/\mathfrak{M}_1$. It is true that in deriving this result Kuiper had made use of the mass-luminosity relation, and there is now considerably more doubt whether this relation may be applied indiscriminately to all the components of binary stars. Nevertheless we must accept his verdict that there is no obvious separation between the close pairs and the wide pairs.

Next, Kuiper has shown that the frequency-curves of the semimajor axes leave little doubt that the binaries of a given absolute magnitude and spectral class form one kind of objects and not two groups. He also pointed out that fission cannot take place in stars which are built in accordance with Eddington's standard model, or even in stars which are considerably less concentrated toward the center than the standard model. Moreover, fission cannot account for the origin of binaries which are wider than a few astronomical units. The maximum frequency of the binaries is for separations of 15 to 20 astronomical units.

The fission theory would lead us to expect that the frequency-curves of the mass ratios for various groups of main-sequence stars would be the same, because the internal constitution of a main-sequence star would be about the same no matter where on the sequence it is located. Hence, if fission is the result of a single mechanism such as rotational instability, then the dividing parts would always have the same mass ratio. But Kuiper's work has not confirmed this conclusion. According to him, stars with total masses of the order of twenty times that of the sun have mass ratios $\mathfrak{M}_2/\mathfrak{M}_1 = 1/\alpha$ which increase in frequency as we go from $1/\alpha = 1.0$ to $1/\alpha = 0.0$. In other words, there is a tendency among the very massive stars of the main sequence to have companions of relatively small mass. For stars with total masses of four times that of the sun or less, the distribution is different. In these systems the ratio is somewhat more often in the vicinity of $\alpha = 1.0$ than for smaller values. This shows that if the primary component of a binary system located on the main sequence is a star of moderate mass, the companion is quite likely to have a mass only a little smaller than the primary. Kuiper has generalized his conclusions in the following manner. Given a certain primary, it is almost a matter of chance what the mass of the companion may be. The only factor which comes into play is the luminosity function (which may not be identical with that of the single stars), so that if we have, for example, a primary of very large mass, then because of the great frequency of stars of relatively small mass per unit volume, the chances are considerable that the companion will be an object of small mass; but if the primary is itself a star of approximately the solar mass, then the chances are very good that the companion will also be a star of, roughly, the solar mass.

Before we drop this subject it might be well to recall one other relation which militates strongly against any hypothesis that would attribute a different origin to the close pairs and the wide pairs. This is the well-known relation between the period of a double star and the eccentricity of its orbit. In his Darwin lecture before the Royal Astronomical Society in 1932, R. G. Aitken presented the data for spectroscopic binaries as well as for visual binaries (Table XII).

Aitken stated that this is perhaps the only relation among the binary stars which has been established beyond reasonable doubt. It is also probable that it has a physical basis and is in some way related to the origin and evolution of binary and multiple systems. He says: "Unquestionably the eccentricity, on the average, increases with the period; the spectroscopic orbits of shortest period are very nearly circular, the visual orbits of longest period are greatly elongated ellipses."

When the two components of a visual binary are of approximately equal magnitude, they are usually indistinguishable in color. Aitken had noticed this, particularly among the hundreds of pairs with angular distances not exceeding 0.3″. As a rule, he found that in at least nine cases out of every ten, the difference is less than 0.5 magnitude, and there is no appreciable difference in color. This result has been further strengthened by F. C. Leonard, who found that when the two components of a visual binary are equal or nearly equal in brightness, their spectra are almost always approximately the same. For systems with greater differences in apparent magnitude, the dif-

TABLE XII

Distribution of Binary Stars According to Eccentricity

Number	Average Period	Average e
	Spectroscopic Binaries	
83	2.7 days	0.05
49	7.6 days	0.16
29	14.1 days	0.22
23	30.6 days	0.35
21	102.5 days	0.30
31	1177 days = 3.2 years	0.31
	Visual Binaries	
14	6.8 years	0.43
24	37.1 years	0.40
24	73.0 years	0.53
23	138.0 years	0.57
18	274.3 years	0.62
	Visual Binaries Statistically Considered	
500	2,000 years	0.61
800	5,000 years	0.76

ference in spectral class is also considerably greater, and the general rule is that the fainter component is of later type and is therefore redder. There are, however, a large number of pairs for which this is not true, the fainter component being of earlier type, i.e. bluer than the primary. This distinction between the two classes of systems appears to be one of absolute magnitude. When the components of a visual binary are both stars of the main sequence, the brighter star is usually that of earlier type and the fainter star is that of later type. In those binaries in which the fainter component is of the earlier spectral type, the primary is always a late-type giant.

This kind of system, consisting of a late-type giant and an early-type fainter primary, is often of great interest. There are some anomalous stars of this kind; for example, Antares or Alpha Scorpii.

Its primary is a supergiant of class M. Its secondary is an abnormally faint B-type star with emission lines of [Fe II] (p. 92).

The distribution of the binary stars and the relationships among them leave little doubt that they are all formations of the same kind. Although this might be considered to be a serious obstacle against the fission theory, it is not certain that we must necessarily abandon it. The question of the evolution of binary stars has not been entirely exhausted. The work of Moulton and others has shown that at the present time stellar encounters are not sufficiently frequent to explain the large number of visual systems with large semi-major axes. But the lifetimes of most stars on the main sequence are comparable with the lifetime of the galaxy as a whole. It is probable that when most of the visual binary stars were formed, the galaxy was not as it is now. It may have been much more condensed, in which case stellar encounters would have played a more important role than they do now. Furthermore, the question arises whether sources of internal energy in the stars may not have produced a greater separation of the components than is possible from the consideration of purely mechanical causes, such as axial rotation and orbital revolution. After all, we know that the stars possess large stores of internal energy. We have already seen that some of these stores are operating to produce such phenomena as the rings or streams of gaseous matter in the closer spectroscopic pairs. There remains, of course, the question whether the entire store of internal energy of a star would be sufficient to produce a wide pair.

A view that is favorable to the idea of fission was expressed by H. N. Russell in an article in the *Astrophysical Journal* in 1910. In a penetrating analysis of the observed triple and other multiple systems, he showed that their remarkable tendency to consist of one or more close pairs separated from the third or fourth companions by relatively great distances is precisely what might be expected according to the fission theory.

Dr. Russell has permitted me to quote his present views concerning this question, from a letter dated December 18, 1948:

"The statistical result is still good—that the distribution of apparent distances in triple physical systems can be accounted for by supposing that the real separation of the close pair is 9 per cent that of the wide pair in 70 per cent of the cases, and much less in the rest. Also the formal analysis of fission is correct in concluding that, if the separated parts divide without loss of angular momentum, the average ratio of separation of a resulting close pair to that of the wide pair will be of the order of $2.5c^2$, when c depends on the central condensation of the parts after the first separation.

"The observed distribution indicates that for 70 per cent of the systems, the value of c after separation was of the order of 0.19—if the parts lost no angular momentum either by tidal interaction or otherwise into space—and otherwise greater. (Sensible tidal losses now appear very improbable; but other ways of loss have recently been suggested by competent investigators.)

"The relation between the value of c and the ratio $\rho_c/\bar{\rho}$ of the central to the mean density is a definite quantity for any polytropic index n. Curiously enough, I have not been able to lay my hands on any values of c, and have had to calculate rough values by quadrature. They run as follows:

n	0.0	1.0	1.5	2.0	3.0	4.0
c	0.40	0.26	0.21	0.156	0.076	0.024
$\rho_c/\bar{\rho}$	1.00	3.29	6.00	11.4	54.3	62.3
$c(\rho_c/\bar{\rho})^{1/2}$	0.40	0.47	0.51	0.52	0.56	0.60

"It follows that, if the 70 per cent majority of the close pairs in triple systems were produced by fission of such type that just after separation the components were rotating with uniform angular velocity, their central condensation must then have been small, corresponding at most to $\rho_c/\bar{\rho} = 8$.

"This appeared to be a fairly high concentration in 1910, but not today. Theoretical studies and observation of apsidal motions agree in showing that for stars in the upper half of the main sequence which derive their energy from Bethe's cycle $\rho_c/\bar{\rho} \approx 100$. To this would correspond $c \approx 0.06$; $2.5c^2 \approx 0.01$.

"Whether the 'proto-stars' which might have produced ordinary visual or spectroscopic binaries by fission could have had a very much lower central condensation is still unknown, since (so far as I know) no satisfactory theory of the internal constitution of a gaseous mass undergoing a 'Kelvin contraction' has yet been developed, even for internal temperatures high enough to permit the use of Morse's tables of opacity. For lower temperatures the theory of opacity is more difficult, and the possible presence of 'smoke' makes things still worse.

"It is of course not certain that such a proto-star would rotate with uniform angular velocity (compare Jeans' suggestions regarding radiative viscosity).

"Finally, it may well be that the sharp separation of wide and close pairs in multiple systems is a purely dynamical effect, arising from instability when the periods are comparable. This was suggested by Moulton in 1899, but no general theory has yet appeared. It may well be a bad case of the problem of three bodies. No decisive answer can be made until some fairly good theories of this, and of the internal constitution of proto-stars have been developed.

"The great abundance of triple and multiple systems—increased by discoveries of astrometric binaries—makes it certain that they arise from some standard process. But in our present ignorance of the constitution of a star in its early stages, before nuclear energy is tapped, and even of the internal structure and energy supply of giants and supergiants, definite conclusions are premature.

"Such cases as Alpha Geminorum and Epsilon Hydrae strongly suggest repeated fissions—three consecutive ones for the latter."

A very interesting suggestion has recently been made by Unsöld and von Weizsaecker, namely, that the formation of a binary may be the consequence of a supernova outburst. Unsöld, in particular, has stressed the close connection between supernovae and stars of spectral types O and B. He has pointed out that the spatial distribution of supernovae in extra galactic nebulae, according to F. Zwicky, greatly favors the outer regions of these nebulae and avoids the inner and much denser regions. Precisely this same distribution has also been observed in the case of the hotter stars. They, too, seem to be frequent in the outer spiral arms and they do not occur in large numbers in the inner regions of the extra galactic systems. This is an important argument, because if supernovae resulted from some such process as the collision of two ordinary stars, then we would expect them to be much more frequent in the inner regions of the extra-galactic nebulae than in their outskirts.

Zwicky has shown that in each extragalactic nebula there appears one supernova approximately once every 600 years. This agrees well with the estimate made for our own galactic system that during historic times two or three supernovae have appeared, namely the Nova of the year 1054 which is connected with the Crab Nebula, Tycho's Nova of 1572, and perhaps Kepler's Nova of 1604. Unsöld has made an attempt to estimate the frequency of the supernovae. The lifetime of a star of spectral type O is of the order of 10^7 years. That of a main-sequence star of class G0 is 10^{11} years. The number of stars of any given absolute magnitude per cubic parsec, in the vicinity of the sun, is given by the luminosity function which is known from the work of J. C. Kapteyn, P. J. van Rhijn, W. J. Luyten, and others. If we make the assumption that within its lifetime each star produces one super-nova—in other words, if we assume that each star within its process of evolution explodes once as a supernova—we obtain the total number of supernovae within our galaxy per year, if we divide the luminosity function for each absolute magnitude by the lifetime of the star and then take the sum for all absolute magnitudes. In this manner, we obtain 3×10^{-13} supernovae per year per cubic parsec. In order to obtain the entire number of supernovae per year, we must next multiply this quantity by the volume of our galaxy, or 10^9 cubic parsecs. This is based upon the data obtained by Hubble and others concerning the distribution of stars within a typical spiral nebula. The result is that one supernova should appear every three thousand years. This is several times larger than Zwicky's estimate of one supernova in 600 years, but it must be remembered that both this estimate and the computation by Unsöld are based upon rough assumptions. Moreover, Unsöld believes that his value may be in excess by a factor of the order of 10, largely because the lifetime of an average star of a certain absolute magnitude is an upper limit.

At any rate, there is some probability that supernovae actually do have some connection with the stars of early spectral classes. This conclusion has been developed further by von Weizsaecker. He objects to earlier suggestions by Baade and Zwicky, that a supernova is a star which has exhausted its sources of normal energy and is collapsing into the state of a neutron star. He believes that such rapid collapse of matter within the gravitational field of the star would not be possible because of the large rotational momentum inherent in most of these objects. Instead, he believes that the formation of a double star resulting in the outburst of a supernova is a much more likely supposition. The final disruption of the elongated pear-shaped body into two separate objects must have a catastrophic character and must produce the effect of opening suddenly the inner portions of the star, which are then compelled to adjust themselves to the surface conditions of a normal star. In this manner a quantity of energy might well be radiated which would be approximately one order of magnitude smaller than the total thermal content of the O-type star. The latter may be estimated at 5×10^{48} ergs. The resulting radiation of energy would be approximately consistent with what we observe in a real supernova.†

It has been suggested that double stars may have originated from the condensation of separate nuclei in the original cloud of gaseous material. One way to test this hypothesis is to examine the members of galactic clusters for duplicity. Relatively little is known concerning this. Kuiper has examined the Pleiades and has found a considerable number of them to be visual binaries. He concluded, in 1947, that the proportion of visual binaries among them is at least as large as the proportion in ordinary galactic stars, but a study of the radial velocities of the Pleiades by Burke Smith and O. Struve shows that

† Since this was written Dr. von Weizsaecker has informed me that he no longer believes that the binaries have originated from supernovae.

there are very few spectroscopic binaries of large range in radial velocity. Of course, the brighter stars of the Pleiades have very indistinct lines, and small changes in radial velocity would not be noticed. Nevertheless, these stars have been observed probably as much as any other group of stars in the sky. Yet no one has been able to prove that any of the brighter members of the Pleiades are spectroscopic binaries. Among the stars of types A and later, many have sharp lines; it is surprising that no good spectroscopic binary with double lines has been discovered among them. On the other hand, there is strong evidence that spectroscopic binaries are numerous in other clusters. A striking example is NGC 6231. This is not a dense cluster, but it consists of a considerable number of early-type stars with an admixture of several objects of the P Cygni type and of several stars of the Wolf-Rayet type. Some years ago the radial velocities of twenty members of this interesting cluster were measured on McDonald plates. Sixteen of them have normal spectra of types O and early B. Of these sixteen, six have been announced as spectroscopic binaries and one more is probably a spectroscopic binary. This is a large ratio, but it is not yet possible to state definitely that the fraction of spectroscopic binaries to single stars in this cluster is larger than that of a random group of stars of spectral types O to B2. Among the peculiar stars in the cluster there is at least one Wolf-Rayet star which is also a spectroscopic binary. The other Wolf-Rayet stars seem to have a constant radial velocity. So do the other three peculiar stars, two of which are classified as Of.

The existence of numerous spectroscopic binaries in galactic clusters has also been commented upon by Trumpler. For example, in the double cluster of Perseus there are a number of stars known to have variable radial velocities. In several other clusters, for which the color-magnitude diagrams have been accurately determined, there is evidence of a considerable number of double stars. All in all, we must probably conclude that double stars occur in some galactic clusters as often as they do among stars of similar physical characteristics throughout the galaxy. On the other hand, the case of the Pleiades suggests that some clusters may exist without having large numbers of close binaries. The frequency of double stars in galactic clusters is not uniform and is not proportional to the general stardensity in the cluster. Such a proportionality might have been expected if the close binaries had originated from separate nuclei which were formed in the original diffuse cloud. It seems that the origin of double stars is not related to the existence of a cluster, but that the systems have been formed by a process which is independent of the cluster. This would throw doubt upon those theories which attribute their origin and evolution to perturbations caused by neighboring stars. Apparently the density of a stellar system must be much greater than that of a normal galactic cluster in order to appreciably change the equilibrium between binaries and single stars. Ambarzumian has shown that the statistical relations between the elements of binary stars indicate that there is no equilibrium among these objects. This means that the time of relaxation for the binary stars has not yet elapsed, and consequently the age of the great majority of binary systems cannot greatly exceed 10^9 years.

Ambarzumian has also considered the dissociative equilibrium which is produced through the accidental approach of three stars. This investigation was intended to show whether or not binary stars can have resulted as the consequence of accidental triple close passages. It is a comparatively easy mathematical task to write down the appropriate equation of equilibrium. It turns out that the observed ratio of the number of binary stars to the number of single stars is several million times greater than would be expected if there existed a dissociative equilibrium. Hence, at the present time processes of dissociation take place millions of times more often than processes of association. Ambarzumian concludes that the great majority of binary stars in the galaxy cannot have been produced by close approaches. Instead, he believes that the components of each pair have the same origin.

We now know that there is a large range in the values of $\alpha = \mathfrak{M}_1/\mathfrak{M}_2$. It would be unreasonable to find such a range if the double stars had all been produced by the process of classical fission. But if the process of division is catastrophic, all values of α are possible. It would not be surprising that, if the original star was of very early spectral type, the resulting distribution of α would resemble that obtained by Kuiper. For example, if several stars of spectral type O divide into components with mass ratios ranging from 1 to 0.1, then we would obtain a frequency distribution of α, or of Δm, which is essentially determined by the process of division and may be quite uniform. If the original star is of intermediate spectral type, the division into components would favor the discovery of mass ratios of the order of unity, because we would not be able to see or in any other way observe components with masses of the order of $0.1\mathfrak{M}_\odot$, or less.

When we advocate the catastrophic origin of binary stars we really evade all discussion of the physical mechanism involved. We merely believe that there are single stars of appropriate mass and angular momentum which could produce binaries of the types observed in nature. The process might be rapid, or it might involve a number of intermediate stages such as the formation of an explosive envelope as in ordinary novae.

During the past few years a very interesting property of close binary stars has become apparent: there are a considerable number of systems of different types in which the fainter components differ very greatly from the mass-luminosity relation. We have already discussed the problem of the W Ursae Majoris systems in which the spectra of both components are usually visible, despite the fact that on the average the mass ratio $\mathfrak{M}_1/\mathfrak{M}_2$ is approximately equal to 2. This represents so large a difference in mass that if both components were in accordance with the mass-luminosity relation, it would not be possible to observe the spectra of both stars at the same time.

An even more striking case is found among the Algol-type binaries. These systems, as we have already seen, usually show only the spectrum of the smaller, brighter, and hotter component outside of eclipse. The spectrum of the larger, fainter, and cooler component can often be observed during the total eclipse. In a few cases, for example in U Sagittae and U Cephei, the slope of the velocity-curve of the fainter component can be derived during the total phase of the eclipse, and in that case the mass ratio is accurately known. But in the majority of the Algol-type stars this method is not applicable because we cannot get enough observations during the short total phases of the eclipse. Statistical methods show that when α differs greatly from unity, the fainter component is a star which is not often observed elsewhere. From their spectra and their dimensions we would classify them as subgiants. They are not large enough to be regarded as true giants, but they are certainly much larger than ordinary main-sequence stars. Their densities are also low, but their masses are surprisingly small and are not consistent with the mass-luminosity relation. Perhaps these objects are not stable and have not yet come to equilibrium after their formation. In the process of readjusting themselves to the main sequence in the H-R diagram and to the usual mass-luminosity relation, these fainter components liberate an excessive amount of energy which, in one way or another, may affect the entire system. Perhaps it even affects the evolution of the binary stars, increasing their separation and changing their orbits. But what is the fate of those binary stars which we discover spectrographically to have mass ratios close to unity? In these systems the two components are more nearly consistent with the mass-luminosity relation, and we must conclude that a complete or nearly complete adjustment to the main sequence and to the mass-luminosity relation has already taken place. The disturbance is greatest when a small portion of the original star is torn off by the explosion and is compelled to convert itself into a separate star.

It is convenient to trace the probable development of a binary star in the H-R diagram of Figure 8, which shows the lines of equal radius and equal mass. The typical W Ursae Majoris system consists of two components of fairly late spectral type and we may assume that the temperature is somewhere around 5,500°. It is probable that both components lie on the main sequence. Since the mass ratio is, on the average, about 2.0, we must place the two components in slightly different positions on that sequence where it crosses the lines $\mathfrak{M} = 1\mathfrak{M}_\odot$ and $\mathfrak{M} = \frac{1}{2}\mathfrak{M}_\odot$. In accordance with our hypothesis, the W Ursae Majoris system has gradually developed from a binary star of earlier type. In the beginning stages we have a single star lying approximately on the main sequence at $T = 20,000°$ and $\mathfrak{M} = 20\mathfrak{M}_\odot$. If this star splits into two components, we obtain stellar masses both of which are smaller than that of the parent star. If the mass ratio is equal to one, each component has a mass equal to $10\mathfrak{M}_\odot$. The two stars will then fall slightly lower than the parent body, on or near the main sequence. Among the early-type spectroscopic binaries, especially those of type late O and early B, there are some which seem to fall above the main sequence. For example, HD 47129 or AO Cassiopeiae have pronounced giant or supergiant characteristics. They are clearly not stars which are comparable to 10 Lacertae. This may account for the well-known spread in the average densities and sizes of components of early-type spectroscopic binaries. For example, it was pointed out by Pearce that in AO Cassiopeiae the mean densities are much smaller and the dimensions considerably larger than those of Y Cygni. If the star divides into two unequal components, so that $\alpha = 10$, the larger mass differs relatively little from that of the parent star. We locate it slightly below the parent star, on the main sequence. In this case the catastrophe resulting in the splitting of the star will affect the larger mass much less than would be the case if the star breaks into two equal components. The smaller body would now have a mass $\mathfrak{M}_2 = 1.8\mathfrak{M}_\odot$, and this would lower it within the network of the diagram. The spectral type of the companion would then be about A or F. But, since this companion is usually a subgiant, we would have to place it above the main sequence. In the process of adjustment it must slide along a curve of equal mass, but this curve is not the one drawn in Figure 8, because the star does not obey the mass-luminosity relation.

Our hypothesis requires the early-type binary to slide down along the main sequence and finally become a W Ursae Majoris star. Clearly, this evolution involves a large decrease of mass. The only place this mass can go to is the original interstellar medium. If in this process 95 per cent of the original stellar mass is lost and if the age of an early-type rapidly rotating star is 10^3 or 10^4 times less than the age of the galaxy, then there must already have taken place a thorough mixing of the stellar and interstellar material. Most of the interstellar material which we now observe must have been inside a star at some time in the past. Hence, it need not surprise us that B. Strömgren and L. Spitzer, Jr., find the same chemical composition in the interstellar gas that we had obtained in Chapter I for the stars. Even the low abundance of Li and Be in the stars appears to be duplicated in the interstellar gas; Spitzer finds that the ratio Li/Na in interstellar space does not exceed 10^3 times the solar value of 6×10^{-6}, and that the Be/Na ratio in interstellar space is less than about 0.1 of the value for the earth's crust and for the meteorites.

We have shown that the von Weizsaecker process involves not only loss of mass, but also loss of angular momentum. Let us see whether the observed angular momenta in different kinds of stars are consistent with this hypothesis.

We start with a system of the W Ursae Majoris type. The total angular momentum of a close binary with a circular orbit is

$$M = \left(\mathfrak{M}_1 k_1{}^2 + \mathfrak{M}_2 k_2{}^2 + \frac{\mathfrak{M}_1 \mathfrak{M}_2}{\mathfrak{M}_1 + \mathfrak{M}_2} a^2 \right) \omega,$$

where k_1 and k_2 are the radii of gyration of the two stars (cf. p. 127) and a is the distance between their centers. We shall suppose that the central condensation is appreciable, so that

$$k_1{}^2 = k_2{}^2 = 0.3R^2,$$

where $R = R_1 = R_2 = 0.4a$ is the radius of each component. Moreover, $\mathfrak{M}_1/\mathfrak{M}_2 = 2$. Hence,

$$M = 0.4\mathfrak{M}_1 a^2 \omega.$$

If the binary should coalesce to form a single star of mass $\mathfrak{M}_1 + \mathfrak{M}_2 = 1.5\mathfrak{M}_1$, its radius would, according to Figure 8, be about $R' = 1.2R = 0.48a$. The angular momentum of this single star would be

$$M' = \mathfrak{M}' k'^2 \omega' = 0.1\mathfrak{M}_1 a^2 \omega'.$$

Hence, if $M' = M$, we have

$$\omega'/\omega = 4.$$

The linear velocity of the equatorial rotation of each component in the W Ursae Majoris system is about 100 km/sec. We have

$$v_{rot} = \omega R$$

$$v'_{rot} = \omega' R' = 4\omega \times 1.2R = 500 \text{ km/sec.}$$

Such rotational velocities are never observed among single stars of mass $\mathfrak{M}' = 1.5\mathfrak{M}_1 = 1.5\mathfrak{M}_\odot$.

But W. Baade reminds us that only a few hundred single stars of this mass have been observed with slit spectrographs. Since, at the present time, perhaps one star in a thousand of the appropriate class is a W Ursae Majoris system, it is possible that we have not yet observed enough specimens to be sure that there are no suitable descendants among the single stars of classes G, K, and M. But rotational velocities of the order of 500 km/sec could be seen even on objective-prism plates; yet they have never been found. Moreover, the evolution of a close binary is probably a fairly rapid process, so that during the lifetime of the galaxy, between 10^9 and 10^{10} years, several "generations" of binaries have gone through their evolutionary processes. The evolution of single stars of the solar type is slow, being limited to the nuclear transformation of H into He. Hence, there should be an accumulation of "descendants" of the W Ursae Majoris binaries. All in all, we are fairly safe in our conclusion.

The computation of M would have been very similar if we had started with a binary of early spectral class, such as U Cephei or U Coronae Borealis, except that in this case the single star with the combined mass of the binary would have a very reasonable rotational velocity, because values up to 500 km/sec have been observed among single O and B stars. But the angular momentum of an average early-type binary, or rapidly rotating single star, is much larger than that of the W Ursae Majoris system. We can set

$$\mathfrak{M}_B = 10\mathfrak{M}_1$$

$$R_B = 3R.$$

If $k^2{}_B = 0.3R^2{}_B$, we have

$$M_B = \mathfrak{M}_B k^2{}_B \omega_B = 9\mathfrak{M}_1 R v_{B,\text{rot}};$$

while for the W Ursae Majoris binary, we had

$$M = 0.4\mathfrak{M}_1 a^2 \omega = 2.5 R v_{\text{rot}}.$$

If $v_{\text{rot}} = v_{B,\text{rot}} = 100$ km/sec, then

$$M_B/M \approx 3.6,$$

so that in the processes I and H in Figure 41 the star discards not only about 90 per cent of its mass, but also much of its angular momentum.

Returning to the idea that the W Ursae Majoris stars, in the process of their evolution, finally coalesce to form a single star, we have indicated that this would result in the formation of a single star having a large amount of rotational momentum. Since we do not observe single stars of the required spectral type and mass having large rotations, we must assume that the rotational momentum somehow is lost in the process of coalescence.

In this connection it is interesting to recall the hypothesis by von Weizsaecker on the origin of the planetary system. In this hypothesis the assumption is made that a well-developed primitive sun is surrounded by a rotating shell of roughly one-tenth the mass of the central sun, in which each particle is assumed to describe a simple Kepler ellipse about the sun. Von Weizsaecker assumes that the mass of the original sun is approximately equal to the mass of our present sun. This is necessary because we are not cognizant of any process that would rapidly change the mass of the central star through ordinary evolutionary changes. Hence, on his hypothesis the original star must have been approximately similar in spectral type and other characteristics to our present sun. If it was different, it must have been lying somewhere off the main sequence, because the evolutionary process with constant mass runs across the main sequence and not along it. This sun presumably had relatively little rotational momentum, but the shell around it contained a large amount of rotational momentum, presumably about 100 times as much as does the system of the planets in the solar system at the present time. The principal advance made by Ter Haar was to explain how a large part of the rotational momentum of the original sun could be carried away in the process of planet formation. Accordingly, the planets which we now observe possess a mass which is only about 0.01 of the mass of the nebulous disk. The lighter elements, especially hydrogen and helium, were lost by the nebulosity, and they carried off a large fraction of the rotational momentum. Without going into the details of the theory, it is of interest that in ordinary stars of spectral type G we do not observe nebulosities of the kind postulated by von Weizsaecker. It is true that such a nebulosity might not be easily detected. Its extent would be approximately 10^{14} cm in diameter and 10^{13} cm in thickness. This would give an average density of about 10^{-9} gm/cm^3. Such a nebulosity would be too small to be observed visually or by means of direct photographs. But it would almost certainly produce observable effects in the spectrum. The atoms of helium and hydrogen would be excited by the central star and would in all probability produce emission lines. Furthermore, the nebulosity would produce absorption lines of the kind which we are accustomed to observe in the case of extended stellar atmospheres. But these atmospheres are always the consequence of large rotational velocities in the parent star, and they occur predominantly among stars of early spectral class, namely, types B and A. We are not acquainted with shells surrounding stars of the solar type, and if they are a normal stage in the evolution of a single main-sequence object, their life-spans must be short.

If, however, we consider the W Ursae Majoris stars, we have before our very eyes something resembling the nebulosity that is postulated in von Weizsaecker's theory. These systems are already surrounded by common envelopes, envelopes which in many respects resemble the von Weizsaecker nebulosities. Although we do not have any knowledge of the physical processes involved, it is quite reasonable to suppose that as the W Ursae Majoris double star evolves in the direction of forming ultimately a single star, the nebulosity becomes more and more extended and finally produces a nebulous disk of the kind required by the theory. The remaining star would then be one of approximately solar type having a mass not greatly in excess of the mass of the sun, and this star would presumably have only a small amount of rotational momentum. The momentum of the original binary star would be transferred to the nebulosity and would be very largely lost in the process of dissipation of this nebulosity. Only a fraction of the original rotational momentum would be preserved in any planets that may result in the nebula.

In this manner the hypothesis of evolution in binary stars serves to bridge the gap that has always existed between the binary stars and the solar system, and it furnishes us with a large number of examples of approximately the sort of nebulae required in the theory. The existence of a fairly dense original shell, or gaseous disk, is required in the hypothesis of von Weizsaecker; otherwise the condensation of separate planetary bodies would have been impossible because of the strong perturbing influence of the central mass of the sun. On the other hand, if it were assumed that the mass of the sun had been distributed more or less uniformly over the entire disk now occupied by the planetary system, in the process of contraction the mass of the sun would have retained most of the angular momentum, which is contrary to the observational fact that the planets now possess 98 per cent of the angular momentum of the solar system.

If we start again with the total angular momentum of a typical W Ursae Majoris system, in which the smaller mass, $\mathfrak{M}_2 = \frac{1}{2}\mathfrak{M}_1$, is at a distance of approximately R from the center of the more massive star, we can modify the expression for the orbital angular momentum—which constitutes the major part of the whole momentum—by introducing Kepler's third law. This gives, approximately,

$$M = \sqrt{\mathfrak{M}_1 \mathfrak{M}_2} \sqrt{a}.$$

Thus, if the system is converted into one in which there is a planet, like Jupiter, with $\mathfrak{M}_2 = 10^{-3}\mathfrak{M}_1$, we would have to increase a by a factor of 2×10^5 in order to preserve M. This would place the hypothetical planet at a distance of $2 \times 10^5 R$ from the central sun. But in our solar system Jupiter is approximately at a distance of $10^3 R_\odot$. Hence, the total angular momentum of the solar system is about 10 times smaller than that of the W Ursae Majoris system. There is another way of estimating the angular momentum of the solar system.

About 98 per cent of it resides in the orbital motions of the planets, and 2 per cent in the axial rotation of the sun. If all planets were combined with the sun, its rotational velocity would be increased from 2 km/sec to 100 km/sec. But the hypothetical single descendant star resulting from a W Ursae Majoris binary would have an axial velocity of 500 km/sec. The angular momentum of the solar system would be about 5 times smaller than that of the W Ursae Majoris system. The discrepancy between the two estimates is caused by our uncertainty of the law of rotation of the sun. It certainly does not rotate as a solid body. If we had used Fouché's early estimate of the angular momentum of the solar system as being 28 times that of the uniformly rotating sun, then the velocity of the rotation would be only 56 km/sec, if the angular momentum of the planets were absorbed by the sun. The ratio of the limit for instability, 500 km/sec, exceeds this value by a factor of 9. In order to produce a solar system, the von Weizsaecker process would be expected to carry off about 90 per cent of the remaining angular momentum. This agrees fairly well with the theoretical conclusion that 98 per cent may be carried off by the escaping ring of nebulous matter.

Von Weizsaecker has also discussed the further evolution of a star after it has shed its rotational momentum. He had obtained approximately equal lifetimes for the duration of the energy-generating process of Bethe and for the process of shedding of rotational momentum; hence there is no way of telling whether a star first loses its angular momentum, retaining most of its hydrogen intact, or whether it first loses all available hydrogen by converting it into helium without having divested itself of its surplus angular momentum. Perhaps both possibilities are realized in nature. If a star loses its rotational momentum without having been depleted of hydrogen, such an object would remain on the main sequence. The physical parameter which distinguishes a star in different parts of the main sequence is the mass. Since this star is losing mass in the process of shedding its rotational momentum, it is clear that it must move towards the later spectral types. Thus, one would expect fast rotations among the early types and slow rotations among the later types. This is qualitatively in accordance with the observations that rotational velocities are large only in spectral types earlier than about F5. But von Weizsaecker points out that the observations suggest a much more sudden change in the rotational velocities of the stars. This evolution must require a time of the order of the age of the galaxy. We know that the stars of spectral types later than F5 have a relatively small concentration toward the galaxy. These stars are regarded by von Weizsaecker as old stars. The early-type stars, those of types O and B and perhaps A, have a much greater galactic concentration, and are thus young.

The proposed scheme of evolution of close binaries is an extension of the theory of von Weizsaecker, and it accounts for the remarkable fact that among stars of spectral class F5 and later, only the W Ursae Majoris systems have large observed angular momenta. This does not necessarily mean that evolution through the formation of double stars is an important evolutionary path in the H-R diagram. But we see that evolution through loss of mass must somehow occur in the universe on a very large scale, because we cannot otherwise explain the existence of the large number of late-type single dwarfs. We have already seen that single stars of solar type never give observational evidence of the existence of the von Weizsaecker nebula. This fact militates against the acceptance of his process for the formation of the red dwarfs. At the same time, there are many single stars of intermediate spectral types, and they are not accounted for by the double-star hypothesis. Perhaps we must suppose that both processes exist and reinforce each other. If there are now 1,000 single red dwarfs for every W Ursae Majoris binary, the process of evolution designated as D in Figure 41 would have to proceed at a fairly rapid rate— $T = 10^6$ years at the most—in order to account for the accumulation of red dwarfs in the "sink" at the lower end of the main sequence.

The kernel of our hypothesis is the remarkable fact that single stars of low temperature have negligible rotational velocities, while the W Ursae Majoris binaries, although having similar temperatures, possess large angular momenta which are comparable on one side to those of rapidly rotating hot single stars; and on the other side to those of hypothetical planetary systems which need not differ very radically from the solar system.

The hypothesis is an example of a growing trend, among astronomers, to return to the original interpretation of the main sequence in the H-R diagram as an evolutionary path. This trend is especially apparent in the recent work of V. G. Fessenkoff who has discussed the evolution of stars by the loss of mass and angular momentum through the process of "corpuscular radiation," which is identical with our process of loss of matter in expanding envelopes. According to Fessenkoff the sun may have been, a few billion years ago, a hot star of about 10 times its present mass and 18,000 times its present angular momentum (not counting the orbital momenta of the planets).

A. G. Massevich has developed the theory of evolution of a star built on a given model and sustained by the conversion of H into He. The star is constrained to obey the empirical massluminosity relation and loses mass through corpuscular radiation. The result agrees with that of Fessenkoff; the loss of mass is nearly proportional to the luminosity. The present rate of change in mass is quite rapid while the star is more massive than the sun; but it is so insignificant for $\mathfrak{M} = \mathfrak{M}_{\odot}$ that there is no inconsistency with the observations of the sun. But Schatzman has shown that the loss of mass cannot be explained entirely as the result of rotational instability. There must be more powerful eruptive processes.

12. Peculiar Binaries

The proposed hypothesis fails to explain the origin of binaries consisting of giant or supergiant components, such as Epsilon Aurigae, Zeta Aurigae, VV Cephei, etc. For example, the components of Epsilon Aurigae are both supergiants. The smaller and more luminous star has a spectrum of class F and a luminosity which exceeds that of the sun by a factor of 60,000. Its mass is about 40 times that of the sun. The companion of similar mass, whose orbital period is 27 years, is infrared in color and has a surface area which is about 150 times larger than that of the F star, and its density is about 10^{-9} times that of the sun. The enormous energy-production of Epsilon Aurigae shows that it cannot be older than about 10^7 years. If rotational evolution will not change this system, nuclear processes will undoubtedly accomplish it in 10^7 years. The angular momentum of the system is large, and it is not immediately obvious how the star can get rid of it. These giant systems, per unit volume, are exceedingly rare.

More frequent are the late-type dwarf systems which show Algol-type light-curves. Their periods are very short, yet their light-curves show no ellipticity effect. One of them is YY Geminorum, the distant companion of the double star Castor, also designated as Castor C. Its period is 0.81 day. There are two spectra of approximately the same type, dwarf K6. The depths of the two minima are both 0.6 magnitude, and the duration of the eclipse is 2 hours. This star is believed to be the nearest eclipsing variable. It is also one of the intrinsically faintest known to us at the present time. Since both stars are dwarf K stars, there is nothing unusual about the low absolute magnitude. The radial velocity of this system was observed at the Mount Wilson Observatory by A. H. Joy and R. F. Sanford. They found that the spectra of two stars are visible. Both produce sets of absorption lines and both have strong emission lines of Ca II and H. According to the Mount Wilson observers, the radial velocities given by the emission lines are the same as those given by the absorption lines. The semi-amplitude of the velocity-curve of the principal component is 114 km/sec, while the semi-amplitude of the velocity-curve of the secondary is 127 km/sec. The sum of the two orbital radii is 2,700,000 km and the mass functions of the two stars are $\mathfrak{M}_1 \sin^3 i =$

$0.63\mathfrak{M}_\odot$ and $\mathfrak{M}_2 \sin^3 i = 0.57\mathfrak{M}_\odot$. With the help of the light-curve by H. van Gent, the following values have been derived for the absolute dimensions of the system.

$$a_1 + a_2 = 2{,}700{,}000 \text{ km} = 0.018 \text{ astronomical units}$$

$$r_1 = 530{,}000 \text{ km} = 0.76R_\odot$$

$$r_2 = 472{,}000 \text{ km} = 0.68R_\odot$$

$$\mathfrak{M}_1 = 0.63\mathfrak{M}_\odot$$

$$\mathfrak{M}_2 = 0.57\mathfrak{M}_\odot.$$

It may be of interest to recall the various orbits involved in this remarkable system. Castor is a famous visual double star whose components are designated as Alpha-one (α^1) and Alpha-two (α^2) Geminorum. The visual orbit is as yet indeterminate. However, it is certain that the period must be several centuries, the estimates varying between 232 and 1000 years. In addition, there is the red-dwarf companion designated as C, which we have just been discussing. Each of the three stars seen visually is a spectroscopic binary, but only the star C has the spectra of both components recorded. The projected distance between the pair $\alpha^1 + \alpha^2$ and the dwarf binary C is about 1.43×10^{10} km, or about 960 astronomical units. The distance between the two stars α^1 and α^2 is about 76 astronomical units. The separation of the two spectroscopic components of α^2, which is approximately equal to the separation of the two spectroscopic components of α^1, is about 1000 times less, and this is also approximately the separation between the two spectrographic components of Castor C.

The spectrum of Castor C is remarkable because of the presence of the bright lines. According to Joy and Sanford, the relative intensities of the two emission components were not always the same between 1916 and 1926. That component which happens to be receding tends to have the stronger emission lines. The absolute magnitudes of the two components are, on the visual system, 9.6 and 9.8. This places them among the faintest binary stars recorded. During the past year (1948–1949), the spectrum of Castor C has been observed at the McDonald Observatory. The variation in the intensities of the bright lines with the phase has disappeared. At first sight it would seem that this variation was opposite in phase to that which we have observed in the absorption lines of many close spectroscopic binaries. But the emission lines of Ca II in late-type binaries are usually attributed to the tidal bulges of the two components, and not to a large nebulous envelope, or ring. The asymmetry in the emission-line intensities can be formally explained as a displacement of the tidal bulges from the line joining the centers of the two stars. Another feature of interest is the sharpness of the emission lines. Those of Ca II are sharp, with a dispersion of 50 Å/mm. But the hydrogen lines are slightly more diffuse, though not nearly as diffuse as Stark effect would cause them to be at the pressure of the reversing layer of a dwarf star. The pressure in the tidal bulges is, therefore, much lower than the pressure in the reversing layer.

Another dwarf system is UX Ursae Majoris, whose period is only 4 hours and 43 minutes, making it the system with the shortest period on record. The spectrum of UX Ursae Majoris was observed by Kuiper and found to be of type B3. This binary has an apparent magnitude of 12.7 at maximum, and the depth of the principal minimum is 1.0 magnitude. Hence, it is difficult to obtain adequate exposures which are short enough not to blur the spectrum and not to superimpose variations due to the orbital motion within the 4^h43^m period. A spectrograph of very small dispersion—220 Å/mm at λ 3933—was used. Several different groups of spectrograms were obtained with exposures of the order of 40 or 50 minutes. There are interesting periodic changes in the spectrum. The absorption lines are strong, and are not particularly broad or shallow, about one hour before principal mid-eclipse. At this stage the Balmer lines can be seen as far as H15. This would indicate that the spectrum does not resemble that of a white dwarf, in which the hydrogen lines would be extremely broad. They look fairly sharp, and they do not give the impression of excessive surface gravity. Perhaps they are produced by a mass of gas at relatively low pressure on that hemisphere of the primary which is visible when it is receding. From the positive sign of the radial velocity, we conclude that we are actually observing the heavier component. Emission lines are present when the absorption lines are almost invisible. This occurs from about the middle of the eclipse, through the entire stage when the radial velocity is negative and the principal component is approaching. In this stage it is extremely difficult to see any absorption features, but a few traces of the higher members of the Balmer series remained visible in absorption. The total amplitude $2K_1 = 500$ km/sec, which in turn would give $a_1 \sin i = 7 \times 10^6$ km and, for the mass function, $0.3\mathfrak{M}_\odot$. If the two masses are alike, and the inclination is 90°, the two stars would each have one solar mass. When the absorption lines are weak, Hβ can be seen as a rather diffuse bright line of slightly greater intensity than that of the neighboring continuous spectrum. It is believed that this emission line is broadened to such an extent as to indicate motions of the order of close to 1,000 km/sec. These motions, obtained from the emission lines, must not be confused with the motion of orbital revolution. They correspond to irregular motions of the emitting hydrogen atoms within the field of the binary. A. P. Linnell at Harvard found intrinsic variations in the photoelectric light-curve with average intervals of 4 minutes and a mean amplitude of 0.03 magnitude. Our hypothesis does not encompass systems of this kind.

INDEX OF AUTHORS

INDEX OF SUBJECTS

INDEX OF STARS

(When a star appears on several successive pages, usually only the first is listed.)

265